THEORY OF NONLINEAR OPERATORS

Reviewers: J. Kopáček, A. Kufner

Theory of Nonlinear Operators

Proceedings of a Summer School
held in September 1971 at Babylon, Czechoslovakia

ACADEMIA
Publishing House of the
Czechoslovak Academy of Sciences

ACADEMIC PRESS
New York and London

PRAGUE 1973

Published throughout the World with the exception of Socialist countries by
Academic Press, Inc., 111 Fifth Avenue, New York, New York 10003

For Academic Press, I.S.B.N.: 0-12-427650-4
Library of Congress Catalog Card No. 72-86370

Printed in Czechoslovakia

CONTENTS

PREFACE

During the last years, the Mathematical Institute of the Czechoslovak Academy of Sciences has organized summer-schools devoted to non-linear functional analysis and its applications - particularly in the theory of boundary value problems for differential equations.

If the actual situation in the research of solvability of non-linear operator equations is to be appreciated, it can be said that in some directions (e. g. in the theory of monotone operators or in the abstract calculus of variations) this theory tends to its completion. However, this cannot be said about problems constituting the non-linear analogue to the well-known Fredholm theorems and to the spectral theory. This topic has been studied very intensively during the last years among other countries also in Czechoslovakia, which is why the main subject of the summer-school held from 13 to 17 September 1971 at Babylon was the spectral analysis of nonlinear operators; nevertheless other problems connected with the solvability of nonlinear operator equations were considered as well.

The summer-school was attended by more than 70 mathematicians from Czechoslovakia and abroad. The lectures were delivered by

Marion S. Berger, New York, N. Y. (USA),
Melvyn S. Berger, New York, N. Y. (USA),
Reinhard Kluge, Berlin (GDR),
Milan Kučera, Praha (Czechoslovakia),
Arno Langenbach, Berlin (GDR),
Joachim Naumann, Berlin (GDR),
Jindřich Nečas, Praha (Czechoslovakia),
Giovanni Prodi, Pisa (Italy),
Aleš Pultr, Praha (Czechoslovakia),
Vladimír Souček, Praha (Czechoslovakia),
Sergio Spagnolo, Pisa (Italy),

Hans Triebel, Jena (GDR),
Elizbar S. Tsitlanadze, Tbilisi (USSR).

In this volume the texts of almost all lectures delivered during the school are collected. We wish to thank Academia, the Publishing House of the Czechoslovak Academy of Sciences, for their willingness to publish these Proceedings.

Organizing Committee
of the Summer-School

ON THE INVERSION OF SOME DIFFERENTIABLE MAPPINGS WITH SINGULARITIES BETWEEN BANACH SPACES*)

ANTONIO AMBROSETTI, GIOVANNI PRODI,
PISA (ITALY)

Some classical methods for the inversion of nonlinear mappings between Banach spaces use as their starting point the local inversion theorem, the invertibility in the large being proved then by various methods. In this direction it is well-known that P. Levy [1] and R. Caccioppoli [2] obtained very interesting results with many applications.

The purpose of this research is to prove that the basic idea of these methods can be usefully employed even when studying a singular set (i. e. the set where the differential is not invertible) and its image (which we call "critical set"). The case that we treat is the simplest in this direction: namely, the case in which both the singular and the critical set are differentiable manifolds of codimension 1 .

We apply our results to the study of the boundary-value problem

$$\Delta u + f(u) = g \, , \qquad (1)$$

$$u\big|_{\partial\Omega} = 0$$

where Ω is a bounded open set sufficiently smooth, and f is a function which is linearly increasing as the argument tends to $+\infty$ and $-\infty$, but without symmetry (see § 3).

This method gives very exact results on the number of the solutions. It is interesting to observe that for our problem Leray and Schauder's method - at least in its more obvious form - gives no useful result, since the topological degree is zero.

*) The research is supported by CNR (Consiglio Nazionale delle Ricerche), Roma.

§ 1

This section is devoted to some simple purely topological properties concerning the inversion of mappings.

We recall:

1.1. Definition. A mapping $\Phi : X \to Y$ (X, Y topological spaces) is said to be proper if for every compact set $K \subset Y$, the set $\Phi^{-1}(K)$ is compact in X.

1. 2. Definition. Let X and Y be topological spaces. A continuous mapping $\Phi : X \to Y$ is locally invertible in $u_0 \in X$ if there exist a neighborhood U of u_0 and a neighborhood V of $y_0 = \Phi(u_0)$ such that Φ induces a homeomorphism between U and V.

We set $N(y) = \# \Phi^{-1}(\{y\})$ (cardinal number of $\Phi^{-1}(\{y\})$).

1.3. Proposition. Let X and Y be metrisable topological spaces, $\Phi : X \to Y$ a proper, continuous mapping which is locally invertible at every point. Then the function $y \mapsto N(y)$ is finite and locally constant.

The fact that $N(y)$ is finite follows obviously from the fact that $\Phi^{-1}(\{y\})$ is discrete and compact. Moreover it is easy to check that $N(y)$ is both upper and lower semicontinuous: hence it is locally constant.

As a corollary of this proposition, we obtain that if Y is connected, then $N(y)$ is constant. Moreover, if X and Y are arcwise connected and Y is simply connected (that is, every loop in Y is homotopic to a constant), then we get that $N(y) = 1$ and Φ is a global homeomorphism of X onto Y. This is the well-known "Global Inversion Theorem"; it can be proved by a method similar to the classical one which is used for the "Monodromy Theorem" in the theory of analytic functions.

For our purpose it is fundamental to study the set of the points at which the mapping is not locally invertible.

1.4. Definition. Let $\Phi : X \to Y$ be a continuous mapping (X and Y topological spaces). We say that $u \in X$ is a singular point if Φ is not locally invertible at u; $y \in Y$ is said to be a critical point if $y = \Phi(u)$, for some singular point $u \in X$.

We shall speak also, with obvious meaning, of singular set and of critical set. Clearly, the singular set is a closed subset of X .

The following proposition is a trivial consequence of 1.3.

<u>1.5. Proposition.</u> Let X and Y be topological metrisable spaces and $\Phi : X \to Y$ a continuous proper mapping. We denote the singular set by W . Then $N(y)$ is constant on every connected component of $Y \setminus \Phi(W)$.

In order to obtain this proposition from 1.3, it is sufficient to consider the mapping $\Phi : X \setminus \Phi^{-1}\Phi(W) \to Y \setminus \Phi(W)$. It is proper and moreover, it is clear that it is invertible at every point $u \in X \setminus \Phi^{-1}\Phi(W)$.

<u>§ 2</u>

Now we consider differentiable mappings between Banach spaces.

<u>2.1. Definition.</u> Let X and Y be real Banach spaces, Λ an open set of X . We say that $\Phi : \Lambda \to Y$ is a $\mathscr{C}^k$ $(k \geq 1)$ mapping if it is k-times differentiable and if the r-th derivative $\Phi^{(r)}$ $(1 \leq r \leq k)$ is a continuous mapping from Λ to the set of the r-linear mappings of X to Y (with the usual norm). We shall use the symbol $\Phi^{(r)}(u_0)[v_1][v_2]\ldots[v_r]$ to denote the value that this r-linear mapping assumes when it is evaluated for the arguments $(v_1,\ldots,v_r)$, u_0 fixed.

<u>2.2. Definition.</u> We say that the $\mathscr{C}^k$-mapping Φ is locally invertible at $u_0 \in X$ if Φ induces a $\mathscr{C}^k$-diffeomorphism between an open neighborhood U of u_0 and an open neighborhood V of $y_0 = \Phi(u_0)$.

It is known that Φ is locally invertible at u_0 if and only if the linear mapping $\Phi'(u_0) : X \to Y$ is invertible.

The notion of local invertibility which we use here is the translation at a differential level of what we have introduced in the previous section at a purely topological level.

Clearly it is not the same as the previous one; however, from now on we shall use only differentiable mappings and therefore the notion of "local invertibility" will always have the latter meaning. So, for example, we shall say that u_0 is a singular point for Φ if $\Phi'(u_0)$ is not invertible.

2.3. Remark. Obviously all the results proved in the previous section still hold if the meaning of local invertibility and of singular set is that from Definition 2.2.

Now we introduce some notions that we shall use in the study of the singular and critical set of a differentiable mapping.

2.4. Definition. Let X be a Banach space. A set $M \subset X$ is called a $\mathscr{C}^k$-manifold, of codimension 1, if for every point $u_0 \in M$ there exist a neighborhood U of u_0 and a $\mathscr{C}^k$-functional Γ : $U \to R$ such that

(a) $\Gamma'(u_0) \neq 0$,
(b) $M \cap U = \{u : u \in U , \Gamma(u) = 0\}$.

It is easy to prove that a $\mathscr{C}^k$-diffeomorphism transforms a $\mathscr{C}^k$-manifold of codimension 1 to a manifold of the same type. Moreover it is possible, locally, to find a diffeomorphism that transforms such a manifold to a linear manifold of codimension 1 .

It is interesting to see how a smooth manifold of codimension 1 disconnects the space.

2.5. Proposition. Let M be a closed connected $\mathscr{C}^k$-manifold $(k \geq 1)$ of codimension 1 in the Banach space X . Then $X \setminus M$ has at most 2 components.

Proof. Suppose that there are 3 open non-empty, disjoint sets A_1 , A_2 , A_3 such that $X \setminus M = A_1 \cup A_2 \cup A_3$. We remark that since $X \setminus M$ is open, each A_i is open not only relatively to $X \setminus M$, but also to X .

We denote by F_i $(i = 1, 2, 3)$ the boundary of A_i ; they are not empty for if A_i had empty boundary, it would be an open-closed set of X . Obviously we have also $F_i \subset M$.

In virtue of the properties of M , for every $u_0 \in M$ we can find an open neighborhood U of u_0 such that $U \cap (X \setminus M)$ has exactly 2 connected components: then only two of the sets A_i can have a non-empty intersection with U . It follows that $U \cap M$ can be contained at most in two sets F_i .

Moreover, if it is $u_0 \in F_i$ then one of the two connected components of $U \cap (X \setminus M)$ is contained in A_i ; then every point of

$U \cap M$ is a boundary point for A_i, that is, it belongs to F_i. Thus the F_i's are closed-open sets of M. Since, by hypothesis, M is connected, we have $M = F_i$. But this is not consistent with the fact proved above, namely that every point of M belongs at most to 2 sets F_i. Q. e. d.

2.6. Remark. We do not know if, under the hypotheses of Proposition 2.5, the connected components of $X \setminus M$ are always two.

Let us now study the situation, for us especially interesting, that the singular and the critical sets are differentiable manifolds of codimension 1.

2.7. Theorem. Let X and Y be Banach spaces, Λ an open subset of X, $\Phi : \Lambda \to Y$ a mapping of class $\mathcal{C}^k$ with $k \geq 2$.

Suppose that $u_0 \in \Lambda$ is such that:

(I) $\Phi'(u_0)$ has kernel of dimension 1 and image of codimension 1.

(II) If $v_0 \in X$ is a non-zero vector such that $\Phi'(u_0)v_0 = 0$ and γ_0 is a functional on Y such that $\mathrm{Im}(\Phi'(u_0)) = \{z : \langle z,\gamma_0 \rangle = 0\}$ then the linear functional

$$z \mapsto \langle \Phi''(u_0)[z][v_0],\gamma_0 \rangle$$

is not identically zero.

Then the singular set W of Φ is, in a neighborhood of u_0, a $\mathcal{C}^{k-1}$-manifold of codimension 1.

If the condition (II) is replaced by:

(II*) $\langle \Phi''(u_0)[v_0][v_0],\gamma_0 \rangle \neq 0$

then we can find an open neighborhood U of u_0 such that $\Phi(W \cap U)$ is a $\mathcal{C}^{k-1}$-manifold of codimension 1.

The proof of the theorem is based on the following Perturbation Lemma which is well known (it is contained in general results concerning the Fredholm operators). For the reader's convenience, we prefer to prove this lemma completely.

2.8. Lemma. Let $T_0 : X \to Y$ (X, Y Banach spaces) be a linear continuous mapping. We suppose that $\mathrm{Ker}\, T_0$ and $\mathrm{Coker}\, T_0$ both have dimension 1. Then either every linear mapping T near enough to T_0 (in the usual norm) is an isomorphism of X to Y, or it has kernel and cokernel of dimension 1.

Proof of the lemma. Let $\{\lambda v_0\}_{\lambda \in R}$ be the kernel of T_0 and γ_0 a nonzero functional defined on Y such that $\mathrm{Im}(T_0) = \{z : \langle z, \gamma_0 \rangle = 0\}$. Let δ be a functional on X such that $\langle v_0, \delta \rangle = 1$, and s an element of Y such that $\langle s, \gamma_0 \rangle = 1$.

The mapping $p : u \mapsto u - v_0 \langle u, \delta \rangle$ is a projector in X; let $\hat{X}$ be the linear invariant subspace (obviously of codimension 1) related to p. In the same way, the mapping $\pi : y \mapsto y - s\langle y, \gamma_0 \rangle$ is a projector in Y; the invariant subspace related to it, $\hat{Y}$, is equal to $\mathrm{Im}(T_0)$. Let π' be the conjugate projector of π.

Let $T : X \to Y$ be any linear mapping; in order to study whether T is invertible, we put $T = T_0 + S$ and $v = \lambda v_0 + w$, with $w \in \hat{X}$. We have to study the equation

$$T_0 w + S(\lambda v_0 + w) = g \tag{2}$$

with g given in Y. Applying to (2) the projectors π and π', we have the equivalent system

$$\begin{aligned} T_0 w + \pi S(\lambda v_0 + w) &= \pi g , \\ \pi' S(\lambda v_0 + w) &= \pi' g . \end{aligned} \tag{3}$$

Since the operator T_0 is an isomorphism of $\hat{X}$ onto $\hat{Y}$, for S small enough, so is also $T_0 + \pi S$. We put $A = (T_0 + \pi S)^{-1}$.

Then the first equation of (3) is equivalent to

$$w = A(\pi g - \pi S \lambda v_0) . \tag{4}$$

Substituting to the second equation, we have:

$$\pi' S(\lambda v_0 - A\pi S \lambda v_0) = \pi' g - \pi' S A \pi g ,$$

that is

$$\lambda \langle S v_0 - S A \pi S v_0, \gamma_0 \rangle = \langle g - S A \pi g, \gamma_0 \rangle .$$

Now, we distinguish two cases:

(a) $\langle S v_0 - S A \pi S v_0, \gamma_0 \rangle \neq 0$ (that is, the vector $S v_0 - S A \pi S v_0$ does not belong to $\mathrm{Im}\, T_0$).

In this case, for each $g \in Y$, the system (3) has a unique solution and hence $T : X \to Y$ is an isomorphism.

(b) $\langle S v_0 - S A \pi S v_0, \gamma_0 \rangle = 0$.

In this case, in order to obtain a solution, it must be $\langle g - S A \pi g, \gamma_0 \rangle = 0$.

We note that, for S small enough, the functional $g \mapsto \langle g - SA\pi g, \gamma_0 \rangle$ is different from zero. Then λ can assume arbitrary values, and the kernel of T has dimension 1.

2.9. Remark. In the case (b) we have $\lambda \neq 0$ for the proper solutions of the homogeneous system. Hence to represent the kernel of T we can set $\lambda = 1$. From (4), for $g = 0$, we obtain

$$\|w\| \leq \|A\|\|\pi\|\|S\|\|v_0\| .$$

This relation implies that Ker T, even in case (b), can be associated with the vector $v_0 + w$, with $v_0 \neq 0$, constant, and w which tends to zero as T tends to T_0.

Proof of Theorem 2.7. We put $T_0 = \Phi'(u_0)$, $T = \Phi'(u)$; the meaning of the other symbols is the same as in the previous lemma.

We set: $S(u) = \Phi'(u) - \Phi'(u_0)$ and we denote by $A(u)$ the inverse mapping of $\Phi'(u_0) + \pi S(u)$ (as mapping of $\hat{X}$ to $\hat{Y}$). It is important to remark that $u \mapsto A(u)$ is of class $\mathscr{C}^{k-1}$. A point u belongs to W if and only if the case (b) occurs. We put

$$B(u) = (\Phi'(u) - \Phi'(u_0)) - (\Phi'(u) - \Phi'(u_0))A(u)\pi(\Phi'(u) - \Phi'(u_0))$$

then condition (b) yields

$$\langle B(u)v_0, \gamma_0 \rangle = 0 .$$

In order to prove that this relation is the equation of a $\mathscr{C}^{k-1}$-manifold of codimension 1, we observe that $B(u)$ is of class $\mathscr{C}^{k-1}$ and that the differential of the functional

$$u \mapsto \langle B(u)v_0, \gamma_0 \rangle$$

evaluated at the point u_0 is given by

$$z \mapsto \langle \Phi''(u_0)[z][v_0], \gamma_0 \rangle .$$

This linear functional, by hypothesis (II), is different from zero. Thus we have proved the first statement of Theorem 2.7.

To prove the second statement, we show that, under hypothesis (II*), we can construct in a neighborhood of u_0 a diffeomorphism which maps W to $\Phi(W)$. (Obviously, this diffeomorphism cannot be Φ itself, since u_0 is a singular point for Φ !)

Namely, we consider the mapping $\Psi : U \to Y$ (where U is a suitable neighborhood of u_0) defined by

$$u \mapsto \Psi(u) = \Phi(u) + s\langle B(u)v_0, \gamma_0\rangle \; .$$

Since the functional $u \mapsto \langle B(u)v_0, \gamma_0\rangle$ is zero on W, Ψ is equal to Φ on W. The differential of Ψ at the point u_0 is

$$z \mapsto \Phi'(u_0)z + s\langle \Phi''(u_0)[z][v_0], \gamma_0\rangle \; .$$

An easy computation shows that $\Psi'(u_0)$ is invertible if and only if

$$\langle \Phi''(u_0)[v_0][v_0], \gamma_0\rangle \neq 0 \; ,$$

which is exactly condition (II*). Hence, if this hypothesis holds, the critical set is - locally - the image of the singular manifold under a diffeomorphism of class $\mathscr{C}^{k-1}$: thus it is a $\mathscr{C}^{k-1}$-manifold of codimension 1.

This completes the proof of Theorem 2.7.

<u>2.10. Definition</u>. Given an application of class $\mathscr{C}^k$ with $k \geq 2$, we shall call ordinary singular point a point for which conditions (I) and (II*) hold.

If u_0 is an ordinary singular pdint, then we can compute locally the number of the solutions of the equation $\Phi(u) = y$.

<u>2.11. Theorem</u>. Let $\Phi : \Lambda \to Y$ (Λ open set in a Banach space X, Y Banach space) be a mapping of class $\mathscr{C}^k$ with $k \geq 2$ and $u_0 \in \Lambda$ an ordinary singular point.

Then, denoting by s a vector which is transversal to $\Phi(W)$ at $y_0 = \Phi(u_0)$, there exist a neighborhood U of u_0 and an $\varepsilon \in R$, $\varepsilon \neq 0$ such that

(a) for each $y \in \,]y_0, y_0 + \varepsilon s]$ the equation $\Phi(u) = y$ has 2 solutions in U,

(b) for each $y \in \,]y_0, y_0 - \varepsilon s]$ the equation $\Phi(u) = y$ has no solution in U.

<u>Proof</u>. Since u_0 is an ordinary singular point, it holds that if U is a suitable neighborhood of u_0, $\Phi(W \cap U)$ is, in a neighborhood of $y_0 = \Phi(u_0)$, a $\mathscr{C}^{k-1}$-manifold of codimension 1.

Using the notation of the previous theorems, we denote by s a vector which is transversal to $\Phi(W)$ at the point y_0 and set $y = y_0 + \eta s$, $\eta \in R$.

We can assume, without loss of generality, $\langle s,\gamma_0\rangle = 1$, $u_0 = 0$ and $y_0 = \Phi(u_0) = 0$. We put

$$r(u) = \Phi(u) - \Phi'(0)u ,$$

$$u = \lambda v_0 + w \quad (\text{with } w \in \hat{X}).$$

The equation $\Phi(u) = y$ yields

$$\Phi'(0)w + r(\lambda v_0 + w) = \eta s .$$

We transform this equation into a system using the projectors π and π' .

We have

$$(5) \qquad \Phi'(0)w + \pi r(\lambda v_0 + w) = 0 ,$$

$$\pi' r(\lambda v_0 + w) = \eta s .$$

Since the operator $\Phi'(0)$ is invertible between $\hat{X}$ and $\hat{Y}$ and $r'(0) = 0$, we obtain from the first equation of (5), in virtue of "Dini's Theorem":

$$w = \sigma(\lambda)$$

where σ is a function of class $\mathscr{C}^k$, defined in a neighborhood of $0 \in R$ and such that $\sigma'(0) = 0$.

Hence the system is reduced to the following equation:

$$\pi' r(\lambda v_0 + \sigma(\lambda)) = \eta s ,$$

that is

$$\langle r(\lambda v_0 + \sigma(\lambda)),\gamma_0\rangle = \eta ;$$

this latter equation can be studied by the following lemma, which is elementary:

<u>2.12. Lemma</u>. Let φ be a real function of class $\mathscr{C}^2$ defined in a neighborhood of $0 \in R$, such that $\varphi(0) = 0$, $\varphi'(0) = 0$, $\varphi''(0) > 0$.

Then there exist two positive numbers ε , τ such that:

(a) for each $\eta \in]0,\varepsilon]$ the equation $\varphi(\lambda) = \eta$ has two solutions, of opposite sign, in $[-\tau,+\tau]$;

(b) for each $\eta \in [-\varepsilon, 0[$ the equation $\varphi(\lambda) = \eta$ has no solution in $[-\tau, +\tau]$.

Proof of Theorem 2.11 completed. We set $\varphi(\lambda) = \langle r(\lambda v_0 + \sigma(\lambda)), \gamma_0 \rangle$.

We remark that it yields $\varphi(0) = \varphi'(0) = 0$, and $\varphi''(0) \neq 0$. Then by Lemma 2.12, we obtain the desired statement, q. e. d.

§ 3

Now we shall study a non-linear problem, making use of the general arguments which we have developed in the previous sections.

Let Ω be an open bounded connected subset of R^N, $\partial\Omega$ its boundary and $\bar{\Omega} = \Omega \cup \partial\Omega$ its closure.

The following notation will be used:

$\mathscr{C}^k(\bar{\Omega})$ will denote the space of the functions which are k-times continuously differentiable on Ω and such that the derivatives can be extended continuously on $\partial\Omega$. With the usual norm

$$\|u\|_k = \sup_{0 \leq r \leq k} \sup_{x \in \Omega} |D^r u(x)| ,$$

$\mathscr{C}^k(\bar{\Omega})$ is a Banach space.

$\mathscr{C}^{k,\alpha}(\bar{\Omega})$ $(0 < \alpha < 1)$ will denote the space of the functions $u \in \mathscr{C}^k(\bar{\Omega})$ such that their k-th derivatives are Hölder-continuous with the exponent α in $\bar{\Omega}$. $\mathscr{C}^{k,\alpha}(\bar{\Omega})$ is a Banach space with the norm

$$\|u\|_{k,\alpha} = \|u\|_k + \sup_{x \neq y} \frac{|D^k u(x) - D^k u(y)|}{|x - y|^\alpha} .$$

$\mathscr{C}_0^{k,\alpha}(\bar{\Omega})$ will denote the subspace of $\mathscr{C}^{k,\alpha}(\bar{\Omega})$ consisting of the functions which are zero on $\partial\Omega$.

$L^p(\Omega)$ will denote the space of measurable functions u such that $|u|^p$ are integrable, with the usual norm

$$\|u\|_{L^p(\Omega)} = \left\{\int_\Omega |u|^p \, dx\right\}^{1/p} ;$$

$L^\infty(\Omega)$ will denote the space of measurable essentially bounded functions, with the ess sup norm.

We shall say that Ω is of class $\mathscr{C}^{k,\alpha}$ if its boundary has, in the neighborhood of every point, a regular parametrization of class $\mathscr{C}^{k,\alpha}$.

Finally we recall that the classical problem:

$$\Delta u + \lambda u = 0 \quad \text{on} \quad \Omega ,$$

$$u\big|_{\partial\Omega} = 0$$

has countably infinite many eigenvalues $\{\lambda_n\}$, arranged according to increasing magnitude and considering their respective multiplicity. The least eigenvalue is simple: thus we have $0 < \lambda_1 < \lambda_2 \leq \leq \lambda_3 \leq \dots$.

Let $f : R \to R$ be a function of class $\mathscr{C}^2$ satisfying the following conditions:

(i) $f(0) = 0$,

(ii) $f''(t) > 0$ for each t ,

(iii) $\lim_{t\to-\infty} f'(t) = \ell'$ with $0 < \ell' < \lambda_1$,

(iv) $\lim_{t\to+\infty} f'(t) = \ell''$ with $\lambda_1 < \ell'' < \lambda_2$.

In what follows α shall be a fixed number in the interval $]0,1]$.

3.1. Theorem. Let $\Omega \subset R^N$ be a bounded connected open set of class $\mathscr{C}^{2,\alpha}$.

Assume that the real function f has the properties (i), (ii), (iii), (iv).

Consider the boundary-value problem

$$\Delta u + f(u) = g \quad \text{on} \quad \Omega , \tag{1}$$

$$u\big|_{\partial\Omega} = 0 ,$$

where g is given in $\mathscr{C}^{0,\alpha}(\bar{\Omega})$ and the solution u is looked for in $\mathscr{C}_0^{2,\alpha}(\bar{\Omega})$.

Then there exists in $\mathscr{C}^{0,\alpha}(\bar{\Omega})$ a closed connected $\mathscr{C}^1$-manifold M of codimension 1 , such that $\mathscr{C}^{0,\alpha}(\bar{\Omega}) \setminus M$ consists exactly of 2 connected components A_1 , A_2 with the following properties:

(a) if $g \in A_1$ then the problem (1) has no solution;

(b) if $g \in A_2$ then the problem (1) has exactly 2 solutions.

Moreover if $g \in M$ then the problem (1) has a unique solution.

Proof. Consider the mapping $\Phi : \mathscr{C}_0^{2,\alpha}(\bar{\Omega}) \to \mathscr{C}^{0,\alpha}(\bar{\Omega})$ defined by

$$\Phi(u) = \Delta u + f(u) .$$

From the hypotheses on f *) it follows that Φ is of class $\mathscr{C}^2$. The differential of Φ evaluated on $u \in \mathscr{C}_0^{2,\alpha}(\bar{\Omega})$ is given by

$$\Phi'(u) : v \mapsto \Delta v + f'(u)v .$$

To complete the proof of Theorem 3.1 we state here some lemmas, which we shall prove in the following sections.

Lemma A. The mapping Φ is proper.

Lemma B. The singular set W of Φ is non-empty, closed and connected. Every point of W is an ordinary singular point.

Lemma C. If $g \in \Phi(W)$, then problem (1) has a unique solution.

Proof of Theorem 3.1 completed. First we study the properties of the critical set $\Phi(W)$. By Lemma A Φ is proper and hence, since W is closed and connected, $\Phi(W)$ is also closed and connected.

We observe that, by Lemmas A and C, Φ induces a homeomorphism between W and $\Phi(W)$.

Since all the points of W are ordinary, Theorem 2.7 implies that $\Phi(W)$ is a manifold of codimension 1. Thus, by Proposition 2.5 we can say that $\mathscr{C}^{0,\alpha}(\bar{\Omega}) \setminus \Phi(W)$ has at most 2 connected components; moreover since Φ is proper then by Proposition 1.5 we get that the number of the solutions of $\Phi(u) = g$ is constant provided g belongs to the same connected component.

To compute this number, we first observe that, for every neighborhood U of u_0, there exists a neighborhood V of $g_0 = \Phi(u_0)$ such that $\Phi^{-1}(V) \subset U$. Otherwise, there should exist an open neighborhood U^* of u_0 and a sequence u_n such that $u_n \notin U^*$ and $\lim_{n\to+\infty} \Phi(u_n) = g_0$. Since Φ is proper, we might extract a subsequence converging to a point u^* such that $u^* \notin U^*$ and $\Phi(u^*) = g_0$, $u^* \neq u_0$: this would contradict Lemma C.

On the other hand, since u_0 is an ordinary singular point, by Theorem 2.11 we can compute locally the number of the solutions of the equation $\Phi(u) = g$ when g lies on a segment which is transversal to $\Phi(W)$ in g_0. These solutions are 2 or 0 according to the

*) Concerning the mapping $u \mapsto f(u)$, we can factorize it in this way: $\mathscr{C}_0^{2,\alpha}(\bar{\Omega}) \to \mathscr{C}^2(\bar{\Omega}) \xrightarrow{f} \mathscr{C}^2(\bar{\Omega}) \to \mathscr{C}^{0,\alpha}(\bar{\Omega})$, where the first and last mappings are inclusions.

side of $\Phi(W)$ on which g lies. Hence (a) and (b) of the theorem are proved.

Finally, if $g \in \Phi(W)$, then by Lemma C the solution of $\Phi(u) = g$ is unique; the proof of Theorem 3.1 is so completed, q. e. d.

§ 4

This section is devoted to the proof of Lemma A. We will recall some propositions from the "Potential Theory" and the "Eigenvalue Theory".

4.1. Proposition. Let $\Omega \subset R^N$ be a bounded set of class $\mathscr{C}^{2,\alpha}$. Let v be the solution of the boundary-value problem

$$\Delta v = h \quad \text{on} \quad \Omega ,$$

$$v\big|_{\partial\Omega} = 0$$

where h is a bounded function. Then, for every fixed α $(0 < \alpha < 1)$ the following estimate holds

$$\|v\|_{1,\alpha} \leq k_\alpha \|h\|_{L^\infty}$$

where k_α is a suitable constant.

Now we consider the following eigenvalue problem, which is more general than the one we recalled in the previous section:

$$\Delta v + \mu \varrho v = 0 \quad \text{on} \quad \Omega ,$$

$$v\big|_{\partial\Omega} = 0$$

where ϱ is a measurable function bounded by two positive constants; we consider generalized solutions; Ω is a bounded connected open subset of R^N. Then:

4.2. Proposition. The eigenvalues are all positive and form a nondecreasing sequence tending to $+\infty$:

$$0 < \mu_1 \leq \mu_2 \leq \mu_3 \leq \dots \leq \mu_2 \leq \dots .$$

(We suppose every eigenvalue is repeated as many times as its multiplicity.)

For the proof: [3], Chap. VI, § 1.

4.3. Proposition. The first eigenvalue is simple (hence it is $\mu_1 < \mu_2$); the corresponding eigenfunction does not vanish on Ω and its values are all of the same sign.

This proposition is contained in a general theorem concerning the nodes of an eigenfunction [3] (p. 452).

4.4. Proposition. The r-th eigenvalue μ_r is a monotone non-increasing function of the coefficient ϱ. Moreover if $\varrho_1(x) < \varrho_2(x)$ a. e., then, denoting by μ_r^1 and μ_r^2 the r-th eigenvalue of $\varrho = \varrho_1$, $\varrho = \varrho_2$ respectively, we have $\mu_r^1 > \mu_r^2$.

4.5. Proposition. There exists a real number $p > 1$ such that the r-th eigenvalue μ_r depends continuously on ϱ in the topology of $L^p(\Omega)$.

These propositions are an obvious consequence of the variational representation of μ_r ([3] p. 406).

In order to prove Lemma A, we find an a priori estimate for the solutions of problem (1).

4.6. Lemma. Let u_n be a sequence in $\mathscr{C}_0^{2,\alpha}(\bar{\Omega})$ and $\Phi(u_n) = \Delta u_n + f(u_n) = g_n$. If the sequence g_n is bounded in $\mathscr{C}^{0,\alpha}(\bar{\Omega})$, then the sequence u_n is bounded in $\mathscr{C}_0^{0,\alpha}(\bar{\Omega})$.

Proof of Lemma 4.6. Suppose that the opposite holds: namely, $\lim_{n\to+\infty} \|u_n\|_{0,\alpha} = +\infty$. We set $z_n = u_n/\|u_n\|_{0,\alpha}$; we have $z_n \in \mathscr{C}_0^{2,\alpha}(\bar{\Omega})$ and $\|z_n\|_{0,\alpha} = 1$. We introduce the real function h defined as follows:

$$h(t) = \frac{f(t)}{t} \quad \text{for} \quad t \neq 0,$$

$$= f'(0) \quad \text{for} \quad t = 0.$$

In virtue of the hypotheses on f, h is of class $\mathscr{C}^1$ and is bounded.

The relation $\Delta u_n + f(u_n) = g_n$ divided by $\|u_n\|_{0,\alpha}$ yields

$$\Delta z_n + h(u_n)z_n = \frac{g_n}{\|u_n\|_{0,\alpha}} . \tag{6}$$

The sequence $g_n/\|u_n\|_{0,\alpha} - h(u_n)z_n$ is bounded in $L^\infty(\Omega)$. By Proposition 4.1 we have that $\|z_n\|_{1,\alpha}$ is bounded. Therefore we can extract a subsequence converging in $\mathscr{C}_0^1(\bar{\Omega})$ (thus also in $\mathscr{C}_0^{0,\alpha}(\bar{\Omega})$) to a function z^* ; we remark that, since $\|z_n\|_{0,\alpha} = 1$, it must be $\|z^*\|_{0,\alpha} = 1$, by the continuity of the norm. In particular, it is $z^* \neq 0$.

We write (6) in generalized form:

$$-\int_\Omega \sum_i \frac{\partial z_n}{\partial x_i}\frac{\partial w}{\partial x_i}\,dx + \int_\Omega h(u_n)z_n w\,dx = \int_\Omega \frac{g_n}{\|u_n\|_{0,\alpha}}\,w\,dx \tag{7}$$

for every $w \in \mathscr{D}(\Omega)$ ($\mathscr{D}(\Omega)$ denotes the space of $\mathscr{C}^\infty$ functions having compact support contained in Ω).

We observe that at the points $x \in \Omega$ where we have $z^*(x) < 0$ it is $\lim_{n\to+\infty} u_n(x) = -\infty$ and hence $\lim_{n\to+\infty} h(u_n(x)) = \ell'$; so at the points where we have $z^*(x) > 0$, it results $\lim_{n\to\infty} h(u_n(x)) = \ell''$. Thus if we set

$$\begin{aligned} a(x) &= \ell' && \text{if } z^*(x) < 0 , \\ &= \ell'' && \text{if } z^*(x) > 0 , \\ &= f'(0) && \text{if } z^*(x) = 0 \end{aligned}$$

we have at every point of Ω , $\lim_{n\to\infty} h(u_n(x))z_n(x) = a(x)z^*(x)$.

Taking the limit, from (7) we obtain (by Lebesgue's Theorem)

$$-\int_\Omega \sum_i \frac{\partial z^*}{\partial x_i}\frac{\partial w}{\partial x_i}\,dx + \int_\Omega a z^* w\,dx = 0 \quad \text{for every } w \in \mathscr{D}(\Omega) ;$$

this relation shows that $\mu = 1$ is an eigenvalue of the problem (in generalized sense)

$$\Delta v + \mu a v = 0 \text{ on } \Omega , \qquad v\big|_{\partial\Omega} = 0 . \tag{8}$$

We prove that $\mu = 1$ is the least eigenvalue. If the opposite holds, we have $\mu_r = 1$, with $r \geq 2$. We compare the problem (8) with the problem

(9) $$\Delta v + \mu \lambda_2 v = 0 \quad \text{on} \quad \Omega ,$$

$$v\Big|_{\partial\Omega} = 0 .$$

Since it is $a \leq \ell'' < \lambda_2$, then by Proposition 4.4 we get that the eigenvalues of (9) are strictly less than the corresponding eigenvalues of (8).

But this is impossible, for the problem (9) should have two (at least) eigenvalues less than one, while $\mu = 1$ is obviously the second eigenvalue for (9).

Then $v = z^*$ is the first eigenfunction of (8) with eigenvalue $\mu = 1$. But then, by Proposition 4.3, z^* is of the same sign on the whole Ω . Now, if we suppose $z^*(x) > 0$ on Ω , then the following equation is fulfilled

$$\Delta z^* + \ell'' z^* = 0 , \quad z^*\Big|_{\partial\Omega} = 0$$

which cannot be, since ℓ'' is not an eigenvalue for $-\Delta$; on the other hand, we suppose $z^*(x) < 0$ on Ω , hence we have

$$\Delta z^* + \ell' z^* = 0 , \quad z^*\Big|_{\partial\Omega} = 0 .$$

Also this relation cannot be true, since ℓ' is not an eigenvalue for $-\Delta$; this completes the proof of Lemma 4.6.

Proof of Lemma A. Let u_n be a sequence on $\mathscr{C}_0^{2,\alpha}(\bar{\Omega})$ such that $\Phi(u_n) = \Delta u_n + f(u_n) = g_n$ is convergent in $\mathscr{C}^{0,\alpha}(\bar{\Omega})$. By Lemma 4.6, we know that u_n is bounded in $\mathscr{C}_0^{0,\alpha}(\bar{\Omega})$ and hence $\Delta u_n = g_n - f(u_n)$ is a bounded sequence in $\mathscr{C}^{0,\alpha}(\bar{\Omega})$. But since, under our hypothesis on Ω , the operator Δ is an isomorphism of $\mathscr{C}_0^{2,\alpha}$ onto $\mathscr{C}^{0,\alpha}$ ([4] p. 335), we can say that u_n is a bounded sequence in $\mathscr{C}_0^{2,\alpha}$. Hence we can extract from u_n a subsequence converging in $\mathscr{C}^{0,\alpha}(\bar{\Omega})$; then the equation itself shows that this subsequence converges in $\mathscr{C}_0^{2,\alpha}(\bar{\Omega})$. So Lemma A is completely proved, q. e. d.

§ 5

This section is devoted to the proof of Lemma B.

First we prove that all the points of the singular manifold W are ordinary, namely, hypotheses (I) and (II*) of Theorem 2.7 are fulfilled.

The differential of Φ at a point u_0 is given - in our case - by the mapping $\mathscr{C}_0^{2,\alpha} \to \mathscr{C}^{0,\alpha}$ defined by

$$v \mapsto \Delta v + f'(u_0)v \ .$$

It is known that the point $u_0 \in W$ (singular set) if and only if the problem

$$\Delta v + f'(u_0)v = 0 \ , \qquad v|_{\partial\Omega} = 0 \tag{10}$$

has proper solutions. This relation is equivalent to the statement that $\mu = 1$ is an eigenvalue of $\Delta v + \mu f'(u_0)v = 0$, $v|_{\partial\Omega} = 0$.

By Proposition 4.4, it is the least eigenvalue since $0 < \ell' < f'(u_0(x)) < \ell''$, with $0 < \ell' < \lambda_1 < \ell'' < \lambda_2$.

Hence it is a simple eigenvalue; the kernel of $\Phi'(u_0)$ is associated with a non-zero vector $v_0 \in \mathscr{C}_0^{2,\alpha}(\bar{\Omega})$. It is known that $\operatorname{Im}\Phi'(u_0)$ consists of the elements $g \in \mathscr{C}^{0,\alpha}(\bar{\Omega})$ for which it is $\int_\Omega g(x)v_0(x)\,dx = 0$. Then hypothesis (I) of Theorem 2.7 is satisfied.

The functional γ_0 which is associated with $\operatorname{Im}\Phi'(u_0)$ is

$$z \mapsto \int_\Omega z(x)v_0(x)\,dx \ .$$

Now we compute Φ'' ; since the second differential of the linear term vanishes, we have

$$(\Phi''(u_0)[v][w])(x) = f''(u_0(x))v(x)w(x) \ .$$

Then condition (II*) of Theorem 2.7 becomes

$$\int_\Omega f''(u_0)v_0^3\,dx \neq 0 \ .$$

This condition is satisfied, since $f''(t) > 0$ for each t and v_0 is of the same sign on the whole Ω , being the first eigenfunction of (10).

To complete the proof of Lemma B, we must show that W is non-empty and connected. We show that W has a Cartesian representation on a linear subspace of $\mathscr{C}_0^{2,\alpha}(\bar{\Omega})$ of codimension 1.

Indeed, let $s \in \mathscr{C}_0^{2,\alpha}(\bar{\Omega})$ with $s(x) > 0$ for every $x \in \Omega$ and let Z be any linear subspace of $\mathscr{C}_0^{2,\alpha}(\bar{\Omega})$ of codimension 1, such that $s \notin Z$. Every element $u \in \mathscr{C}_0^{2,\alpha}(\bar{\Omega})$ can be represented, of course, in a unique way in the form $u = z + \nu s$, $\nu \in R$, $z \in Z$. We consider the eigenvalue problem:

$$\Delta v + \mu f'(z + \nu s)v = 0 \quad \text{on } \Omega ,$$

$$v\Big|_{\partial\Omega} = 0$$

where z is a fixed element of Z, and $\nu \in R$. We denote by $\mu(\nu)$ the first eigenvalue. By Proposition 4.5, μ is a continuous function of ν. Since $s(x) > 0$ on Ω, we have for every $x \in \Omega$:

$$\lim_{\nu\to-\infty} f'(z(x) + \nu s(x)) = \ell' ,$$

$$\lim_{\nu\to+\infty} f'(z(x) + \nu s(x)) = \ell'' .$$

Moreover since $\ell' < f'(t) < \ell''$, we can easily prove that the previous limits are still limits also in L^p norm (for every p). It follows by Proposition 4.5 that

$$\lim_{\nu\to-\infty} \mu(\nu) = \frac{\lambda_1}{\ell'} > 1 ,$$

$$\lim_{\nu\to+\infty} \mu(\nu) = \frac{\lambda_1}{\ell''} < 1 .$$

Thus there exists a value $\bar{\nu}$ such that $\mu(\bar{\nu}) = 1$. This value is unique since μ is a monotone strictly decreasing function (Proposition 4.4).

Hence we have proved that every straight-line $\nu \mapsto z + \nu s$ meets the manifold W in a unique point; it is easy to show that this point depends continuously on z. (Otherwise, we can recall that W is a differentiable manifold and show that the straight-lines $\nu \mapsto z + \nu s$ are transverse to W.) Q. e. d.

§ 6

In this last section we prove Lemma C.

Proof of Lemma C. Let $u_0 \in W$, $\Phi(u_0) = g_0$; we suppose that the equation $\Phi(u) = g$ has another solution $\tilde{u}$.

We set

$$\omega(x) = \frac{f(\tilde{u}(x)) - f(u_0(x))}{\tilde{u}(x) - u_0(x)} \quad \text{if} \quad \tilde{u}(x) \neq u_0(x) \ ,$$

$$= f'(u_0(x)) \quad \text{if} \quad \tilde{u}(x) = u_0(x) \ .$$

Then $\tilde{u} - u_0$ is a proper solution of the problem

$$\Delta v + \mu\omega v = 0 \quad \text{on } \Omega \ ,$$

$$v\big|_{\partial\Omega} = 0$$

with $\mu = 1$. Since it is always $\ell' < \omega(x) < \ell'' < \lambda_2$, we obtain that $\tilde{u} - u_0$ is the first eigenfunction of this problem. By Proposition 4.3 $\tilde{u}(x) - u_0(x)$ has the same sign on the whole Ω and hence in virtue of the hypothesis $f''(t) > 0$ for every t , it follows that $\omega(x) > f'(u_0(x))$ on Ω .

On the other hand, by hypothesis we have $u_0 \in W$; thus also the problem

$$\Delta v + \mu f'(u_0)v = 0 \quad \text{on } \Omega \ ,$$

$$v\big|_{\partial\Omega} = 0$$

has $\mu = 1$ as its first eigenvalue. This contradicts Proposition 4.4 and so Lemma C is proved.

REFERENCES

[1] P. Levy: Sur les fonctions de lignes implicites. Bull. Soc. Math. de France 48 (1920).

[2] R. Caccioppoli: Un principio di inversione per le corrispon-

denze funzionali e sue applicazioni alle equazioni alle derivate parziali. Rend. Acc. Lincei VI, 16 (1932).

[3] R. Courant, D. Hilbert: Methods of Mathematical Physics. Vol. I, New York 1953.

[4] R. Courant, D. Hilbert: Methods of Mathematical Physics. Vol. II, New York 1962.

Istituto Matematico "L. Tonelli", via Derna, 1 - Pisa, Italy

ON NONLINEAR SPECTRAL THEORY

MELVYN S. BERGER, NEW YORK, N. Y. (USA)

In these lectures nonlinear spectral theory is understood as the study of the structure of the solutions of some operator equation $F(x,\lambda) = 0$ (depending explicitly on some parameter λ) as the parameter λ varies. It is convenient to divide problems concerning nonlinear spectral theory into 4 classes:

(i) bifurcation theory (behavior near a point (x_0,λ_0) at which $F(x_0,\lambda_0) = 0$);

(ii) singular perturbation theory (behavior as $\lambda \to \infty$);

(iii) global theory (general behavior for general λ);

(iv) continuation theory (relations between (i) - (iii)).

In §§1 and 2, we show how these problems can be studied in specific instances using some results from abstract nonlinear spectral theory as a guide. In § 3, we sketch the proofs of the new abstract results we use.

§ 1. NONLINEAR SPECTRAL THEORY AND PERIODIC SOLUTIONS FOR AUTONOMOUS DYNAMICAL SYSTEMS

We consider 4 classes of periodicity problems for some typical classical dynamical systems:

(A) Global Problems (in which there is no obvious first approximation for the desired periodic solutions). Determine the structure of the periodic solutions for the system of N ordinary differential equations

$$x_{tt} + f(x,x_t) = 0 .$$

(B) Local Problems (in which there is an obvious first approximation for the desired periodic solutions). Compare the periodic solutions of

$$x_{tt} + Ax + f(x,x_t) = 0$$

with the periodic solutions of the linear system $x_{tt} + Ax = 0$ near $x = 0$ where A is an $N \times N$ constant matrix and $f(x,y) = o(|x| + |y|)$.

(C) Problems of Celestial Mechanics. Again as in (A) and (B) one attempts to study the periodic solutions of systems of the form

$$x_{tt} + \nabla U(x) = 0 ,$$

$$x_{tt} + B(x)x_t + \nabla U(x) = 0 .$$

(D) Problems for "Continuous" Systems. As in (B) one attempts to find the periodic solutions of the mixed problem

$$u_{tt} = \Delta u - f(x,u) , \quad f(x,y) = o(|y|) ,$$

$$u|_{\partial\Omega} = 0 , \quad \Omega \subset R^N$$

near the normal modes of vibration for the associated linearized system.

Classical results for these problems include: For (A); with $N = 2$, $f(x,x_t) = U(x)$, a result of Whittaker - Tonelli.

For (B); with $f(x,x_t)$ real analytic and the positive eigenvalues of A satisfying certain irrationality conditions, a result due to Ljapunov.

Here we show how recent results on nonlinear spectral theory extend these results in the following ways:

(i) The following result extended that of Whittaker - Tonelli to R^N .

Theorem 1 [1]. There exists a one-parameter family of distinct periodic solutions for the system $\ddot{x} + \nabla U(x) = 0$, where $x \in R^N$ and $U(x)$ is a $C^1(R^N)$ real-valued function with

(1) $0 = U(0) \leqq U(x)$ for $x \in R^N$ and

(2) $U(x)$ is convex and $U(x) \to \infty$ as $|x| \to \infty$.

This family is parameterized by the average of $U(x)$ over a period.

(ii) The following result extending the theorem of Ljapunov.

<u>Theorem 2</u> [2]. There exist N one-parameter distinct families of periodic solutions $x_1(R), x_2(R), \ldots, x_N(R)$ (with R sufficiently small) for the system $\ddot{x} + \beta x + \nabla V(x) = 0$, where $x \in R^N$, $\beta > 0$ and $V(x)$ is a $C^2(R^N)$ even function with $|\nabla V(x)| = O(|x|)$. Also as $R \to 0$, the period of $x_i(R) \to 2\pi/\sqrt{\beta}$.

(iii) One can apply these results to the problems of class (C). Consider for example, autonomous conservative perturbations of the Kepler problem defined by the system

$$(*) \qquad \ddot{x} + \frac{x}{|x|} + \nabla U(x) = 0 .$$

In R^2, by a regularization procedure due to Levi-Civita, this system can be transformed to a system of the form described in Theorem 2, with $N = 2$; so that applying Theorem 2, one finds 2 one-parameter families of periodic solutions for (*). Here β is the negative total energy of the perturbed Kepler system.

Furthermore, Theorems 1 and 2 can be applied to obtain approximations to the periodic solutions for problems of class (D).

(iv) We also shall obtain some results on the continuation problem, which asserts that a one-parameter family (existing locally near $x = 0$) can be extended to a <u>global</u> family depending continuously on a real parameter.

(v) Finally, we can compute constructive iteration schemes for the periodic solutions described in (i) - (iv) generalizing the "averaging" procedure of Krylov - Bogoljubov.

§ 2. APPLICATIONS OF NONLINEAR SPECTRAL THEORY TO GLOBAL DIFFERENTIAL GEOMETRY AND QUASILINEAR ELLIPTIC PARTIAL DIFFERENTIAL EQUATIONS

Consider the following problem: (π_N) Let (M,g) denote a smooth compact N-dimensional manifold equipped with some Riemannian metric g. Does M admit a Riemannian metric $\bar{g}$ (conformally equivalent to

g) with arbitrarily prescribed curvature $R(x)$? (Here for $N = 2$, $R(x)$ denotes the Gauss curvature of (M,g) and for $N > 2$, the scalar curvature of (M,g) .)

These problems lead to the necessity of finding solutions for the equations

$$(\#) \qquad \Delta_g u - k(x) + R(x)e^{2u} = 0 \quad (\text{for } N = 2),$$

$$(\#\#) \qquad \Delta_g u - k(x)u + R(x)u^{\beta} = 0 ,$$

where $\beta = (N + 2)/(N - 2)$ for $N > 2$ and Δ_g denotes the Laplace-Beltrami operator on (M,g) .

For $N = 2$, a necessary condition for the solvability of (#) is that the Gauss-Bonnet formula be satisfied (i. e. $2\pi\chi(M) = \int_M R(x)\,dV$, where $\chi(M)$ is the Euler-Poincaré characteristic of M). On the other hand, in (##), we require a solution $u(x) > 0$ on M, and a result of Lichnerowicz shows that $R(x)$ cannot be arbitrarily prescribed.

For (#), the following results have been obtained.

Theorem 3 [3, 4, 5]. If $\chi(M) < 0$, (#) is solvable if $R(x) < 0$. If $\chi(M) = 0$, (#) is solvable if and only if either $R(x)$ changes sign and $\int_M R(x)e^{2v}\,dV < 0$ where $\Delta_g v = k(x)$, or $R(x) \equiv 0$. If $M = S^2$, (#) is solvable if $R(x)$ is positive at some points of S^2 and is even.

This result is obtained by finding the critical points of the functional

$$I(u) = \int_M \left(\frac{1}{2}|\nabla_g u|^2 + k(x)u - R(x)e^{2u}\right) dV .$$

In general, min $I(u)$ over a <u>linear</u> admissible class $\mathfrak{A}$ is infinite, so the desired critical point is of minimax type. Nonlinear spectral theory suggests that we solve the following isoperimetric problem for the solutions of (#):

Minimize $\int_M (|\nabla_g u|^2/2 + k(x)u)\,dV$ subject to the constraint $\int_M R(x)e^{2u}\,dV = 2\pi\chi(M)$ for $\chi(M) \neq 0$ and with the added constraint $\int_M u\,dV = 0$ for $\chi(M) = 0$ provides $\int_M R(x)e^{2v}\,dV < 0$ where v is a solution of $\Delta_g v = R(x)$.

The problem posed by (##) is more difficult since the functional $\int_M R(x)u^{\beta+1}\,dV$ is <u>not</u> weakly continuous in the Sobolev space $\overline{W}_{1,2}(M,g)$ consisting of functions u with mean value zero and $|\nabla u| \in L_2(M,g)$. However we can nonetheless prove

Theorem 4. Suppose $\int_M k(x)\,dV < 0$, then (##) is solvable for any function $R(x)$ with $R(x) < 0$ on M.

Another example of the problems of nonlinear spectral theory involves studying solutions of the Dirichlet problem

$$(*)\qquad \sum_{|\alpha|\leq m} (-1)^{|\alpha|} D^\alpha A(x,u,\ldots,D^m u) =$$

$$= \lambda \sum_{|\alpha|\leq m-1} (-1)^{|\alpha|} D^\alpha B_\alpha(x,u,\ldots,D^{m-1}u) ,$$

$$D^\alpha u\big|_{\partial\Omega} = 0 .$$

Here Ω is a domain in R^N with boundary $\partial\Omega$, and we shall suppose that this system is obtained as the Euler-Lagrange equations for some isoperimetric variational problem: Find the critical points of $\mathcal{B}(u) = \int_\Omega B(x,u,\ldots,D^{m-1}u)$ subject to the constraint $\mathcal{A}(u) = \int_\Omega A(x,u,\ldots,D^m u) = \text{const.}$ over the admissible functions in $\overset{\circ}{W}_{m,p}(\Omega)$. If the functionals $\mathcal{A}$ and $\mathcal{B}$ satisfy certain evenness, positivity, growth, and smoothness hypotheses, and if Ω is bounded, then it is known that (*) possesses a countably infinite number of distinct normalized solutions $(u_n(x),\lambda_n)$ with $\lambda_n \to \infty$ as $n \to \infty$. We shall discuss the following extensions of this result:

(i) removal of the boundedness condition on Ω,

(ii) removal of the positivity condition on $\mathcal{B}$,

(iii) more specialized bifurcation and continuation results provided $\mathcal{A}$ and $\mathcal{B}$ are "nearly quadratic" for $|u|$, $|Du|$, ..., $|D^m u|$ sufficiently small.

Of particular interest in this connection are recent results on the stationary states of the nonlinear wave and Schrödinger equations

$$-iu_t = \mathcal{L}u + f(x,|u|^2)u , \quad x \in R^N ,$$

$$u_{tt} = (\mathcal{L} - m^2)u + f(x,|u|^2)u , \quad x \in R^N .$$

Here a stationary state $u(x,t)$ is a solution of one of these equations of the form $u(x,t) = \exp(i\lambda t)\,v(x)$, where $v(x) \in L_2(R^N)$ and λ are real-valued, and $\mathcal{L}$ denotes a linear elliptic operator of order $2m$. The function $v(x)$ then satisfies an equation of the form

$$(+)\qquad (\mathcal{L} - \beta)v + f(x,|v|^2)v = 0 , \quad x \in R^N .$$

A typical result in this connection is

Theorem 5 [6]. Let $\mathscr{L} = \Delta$, $f(x, |v|^2) = g|v|^\sigma$, and $N > 1$, then (+) has nontrivial solutions in L_2 if and only if either g , $\beta > 0$ and $0 < \sigma < 4/(N-2)$ or $\sigma = 4/(N-2)$ and $\beta = 0$. Furthermore if (+) has nontrivial L_2 solutions, it possesses a countably infinite number of them.

Note that this result is strictly nonlinear, since if $\sigma = 0$ the result is valuous.

The von Karman equations for the buckling of a clamped thin elastic plate are a system of two semilinear elliptic equations together with null Dirichlet boundary conditions. In the Sobolev space $\overset{o}{W}_{2,2}(\Omega)$, these equations can be written in the form $u + Cu = \lambda Lu$. This equation provides an excellent example for the applicability of abstract spectral theory. A new result in this connection is the following

Theorem 6. For $\lambda \in (\lambda_N, \lambda_{N+1})$, the nonlinear von Karman equations have at least N solution pairs $(\pm u_k, \lambda)$ $(k = 1, \ldots, N)$ where $\{\lambda_j\}$ denote the positive eigenvalues of the linearized von Karman equations.

§ 3. SOME GENERAL RESULTS

Here we discuss the new general results on nonlinear spectral theory that we used in the preceding two paragraphs.

Local results

(i) A bifurcation theorem for odd, gradient operators. Consider the operator equation

$$(*) \qquad u = \lambda\{Lu + R(u)\}$$

defined on a real Hilbert space H , where R(u) is an odd, smooth, completely continuous gradient map of higher order near $u = 0$, and

L is a compact self-adjoint linear operator. Suppose for simplicity, that $(Lu,u) > 0$ for $u \neq 0$.

Theorem 7 [7]. If $\dim \ker (I - \lambda_n L) = K > 0$, then not only is λ_n a bifurcation point of (*), but also (*) possesses at least K distinct one-parameter families of solutions $(u_1(R),\lambda_1(R))$, $(u_2(R),\lambda_2(R))$, ..., $(u_K(R),\lambda_K(R))$ where $R = \|u\|^2$ and $\lambda_i(R) \to \lambda_n$ as $R \to 0$.

This result is an extension of a theorem of M. A. Krasnoselskii.

(ii) A convergent iteration scheme for general bifurcation problem. Let X be a Banach space and consider the solutions of the equation

$$(**) \qquad (\beta I - L)u = F(\beta,u,\varepsilon^2 y)$$

where $(\beta I - L)$ is a bounded linear operator mapping $X \to X$ and $F(\beta,0,0) = F_u(\beta,0,0) = 0$. Suppose for $\beta = \beta_0$, Ker $(\beta_0 I - L)$ is nontrivial and $X = \text{Ker}\,(\beta_0 I - L) \oplus \text{Range}\,(\beta_0 I - L)$. Then we define the following iteration scheme for solutions $(x(\varepsilon),\beta(\varepsilon))$ near (u_0,β_0) where u_0 is a normalized element of Ker $(\beta_0 I - L)$. Set $x_0 = \varepsilon u_0$ and $\beta = \beta_0$, then we compute $x_1 = \varepsilon u_0 + v_1$ and β_1 successively from the formulae

$$(I_{N+1}) \qquad (\beta_N I - L)v_{N+1} = P^{\perp}F(\beta_N,x_N,\varepsilon^2 y),$$

$$(\beta_{N+1} I - L)(\varepsilon u_0) = P_0 PF(\beta_N,x_{N+1},\varepsilon^2 y)$$

where P and $P^{\perp}$ are the canonical projections of X onto Ker $(\beta_0 I - L)$ and Range $(\beta_0 I - L)$, respectively, and P_0 is the projection of Ker $(\beta_0 I - L)$ onto $\{u_0\}$.

Theorem 8 [8]. Under general smoothness conditions on $F(\beta,x,y)$ for ε and $\|y\|$ sufficiently small, the iteration scheme (I_{N+1}) is well defined; and if $\dim \text{Ker}\,(\beta_0 I - L) = 1$, then $(x_N(\varepsilon,y),\beta_N(\varepsilon,y))$ converges to a solution $(\bar{x}(\varepsilon,y),\bar{\beta}(\varepsilon,y))$ of (**). A similar result holds if $\dim \text{Ker}\,(\beta_0 I - L) > 1$, provided (**) has a solution $(\hat{x}(\varepsilon,y),\hat{\beta}(\varepsilon,y))$ of the form $\hat{x}(\varepsilon,y) = \varepsilon u_0 + O(|\varepsilon|^2)$; $\hat{\beta}(\varepsilon,y) = \beta_0 + O(|\varepsilon|)$ depending continuously on ε.

Global results

(i) Isoperimetric problems with convex constraints.

Theorem 9. Suppose L is a self-adjoint linear operator defined on a real Hilbert space H with $\dim \operatorname{Ker} L = N > 0$ and $(Lu,u) \geq c\|u\|^2$ for $u \in (\operatorname{Ker} L)^{\perp}$, and $G(u)$ is a strictly convex, smooth, weakly continuous functional defined on H such that $0 = G(0) \leq G(u)$. Then the equation $u = \lambda \nabla G(u)$ has a one-parameter family of nontrivial solutions (u_R, λ_R) with $G(u_R) = R$.

(ii) Nonlinear gradient perturbations of linear eigenvalue problems. We consider the operator equation (defined on a real Hilbert space H)

$$u + Cu = \lambda Lu \tag{+}$$

where L is a compact, positive, self-adjoint linear map and C is a completely continuous, odd, gradient map such that $(Cu,u) > 0$ for $u \neq 0$, and C is homogeneous of degree $p > 1$.

Theorem 10 [9]. Let the eigenvalues of $u = \lambda Lu$ be denoted $\lambda_1 \leq \lambda_2 \leq \lambda_3 \leq \ldots \leq \lambda_N \leq \ldots$ Then for $\lambda_N < \lambda \leq \lambda_{N+1}$, equation (+) has at least N distinct solution pairs $(\pm u_k, \lambda)$ $(k = 1, \ldots, N)$.

Results on continuation

Roughly speaking, the results we discuss are of 3 kinds:

(a) a result using the Ljusternik - Schnirelmann theory for gradient maps [7];

(b) a result based on the degree of a map, extending Krasnoselskii's theorem on bifurcation points at eigenvalue of linear maps of odd multiplicity [10];

(c) a result for complex analytic maps based on J. Schwartz' bifurcation theorem.

These results can be described as attempts to justify a "Principle of Natural Termination" in nonlinear spectral theory which asserts, roughly speaking, that a "bifurcation branch" (x,λ) with respect to $(0,\lambda_1)$ of solutions of $F(x,\lambda) = 0$ can terminate only if

$\lambda \to \infty$, or $\|x\| \to \infty$, or at another bifurcation point of $F(x,\lambda) = 0$ of the form $(0,\lambda_N)$.

BIBLIOGRAPHY

[1] M. S. Berger. Amer. Jour. Math. 1971.

[2] M. S. Berger. Proc. NRL Conf. 1971.

[3] M. S. Berger. Jour. of Diff. Geom. 1971.

[4] F. Warner, J. Kazdan. Unpublished.

[5] J. Moser. Unpublished.

[6] M. S. Berger. Jour. of Funct. Anal. 1971.

[7] M. S. Berger. Proc. AMS Conference on Nonlinear Funct. Analysis 1970.

[8] M. S. Berger, D. Westreich. Unpublished.

[9] J. Hempel. Indiana Jour. Math. 1971.

[10] P. Rabinowitz. Jour. of Funct. Analysis 1971.

[11] J. Schwartz. Comm. Pure Applied Math. 1963.

Belfer Graduate School of Sciences, Yeshiva University, Amsterdam Avenue and 186th Street, New York, N. Y. 10033, USA

FOLGEN UND ITERATIONSVERFAHREN BEI FOLGEN NICHTLINEARER VARIATIONSUNGLEICHUNGEN

REINHARD KLUGE, BERLIN (DDR)

Wir geben hier Folgen und Iterationsverfahren bei Folgen nichtlinearer Variationsungleichungen im Hilbert-Raum in der Form verschiedener Näherungsverfahren zur approximativen Bestimmung von Lösungen nichtlinearer Probleme mit oder ohne Nebenbedingungen im eindeutig wie im mehrdeutig lösbaren Fall an.

A. ALLGEMEINE BEMERKUNGEN

Zunächst führen wir einige konkrete nichtlineare Probleme an, die unter geeigneten Voraussetzungen auf nichtlineare Probleme im Hilbert-Raum führen.

1. Randwertprobleme

Sei $x \in G \subset R_n$, $t = \{t_\alpha : |\alpha| \leq \ell\}$ mit dem Multiindex $\alpha = (\alpha_1, \ldots, \alpha_n)$, und der Zahl der Komponenten $M = M(n,\ell)$, $t(u)(x) \equiv \{D^\alpha u(x) : |\alpha| \leq \ell\}$ für die reellwertige Funktion $u(x) : G \to R_1$ mit den verallgemeinerten Ableitungen $D^\alpha u(x)$ der Ordnung $|\alpha| = \sum_{i=1}^{n} \alpha_i$ und

$$a(u,v) \equiv \sum_{|\alpha|\leq \ell} \int_G A_\alpha(x,t(u)(x))D^\alpha v \, dx$$

mit den Funktionen $A_\alpha : G \times R_M \to R_1$.

Das verallgemeinerte Dirichletsche Problem lautet:

Für $\overline{g} \in L_2(G)$ wird ein $u \in \overset{o}{W}{}_2^\ell(G)$ derart gesucht, daß

$$a(u,v) = \int_G \bar{g}v \, dx \quad \text{für alle} \quad v \in \overset{\circ}{C}{}^\infty(G) .$$

Es ist äquivalent der Operatorgleichung in H

$$(1) \qquad Tu = g$$

mit $H \equiv \overset{\circ}{W}{}_2^\ell(G)$, $(Tu,v) = a(u,v)$ und $(g,v) = \int_G \bar{g}v \, dx$ [18, 5]. Die Lösung dieses Problems wird unter geeigneten Voraussetzungen als Lösung des Dirichletschen Problems

$$(2) \qquad \bar{A}u = \bar{g} \, , \quad D^\beta u|_{dG} = 0 \, , \quad |\beta| \leq \ell - 1 \, ,$$

mit

$$\bar{A}u(x) \equiv \sum_{|\alpha|\leq\ell} (-1)^{|\alpha|} D^\alpha A_\alpha(x,t(u)(x))$$

aufgefaßt. Mit $A = I - t(T - g)$, $t > 0$, ist (1) äquivalent zum Fixpunktproblem

$$(3) \qquad Au = u .$$

2. Variationsprobleme

Gesucht wird eine Lösung des Minimumproblems

$$(4) \qquad f(u) = \min_{v\in H} f(v)$$

mit $H \equiv \overset{\circ}{W}{}_2^\ell(G)$ und $f(v) = \int_G F(x,t(u)(x)) \, dx$.

Für $A_\alpha = \partial F/\partial t_\alpha$ ist $\bar{A}$ der Euler-Lagrangesche Operator von f . Wir erhalten so den Zusammenhang zu (1) und (3). Unter geeigneten Bedingungen gilt: T ist der Gradient von f : $T = f'$.

3. Variationsungleichungen für Differentialoperatoren

Wir betrachten das Problem: Gesucht ist ein $u \in C \subset H$ mit $H \equiv \overset{\circ}{W}{}_2^\ell(G)$ und

$$(5) \qquad a(u,v - u) \geq \int_G \bar{g}(v - u) \, dx \quad \text{für alle} \quad v \in C$$

für die konvexe abgeschlossene Menge C . Eine zu (5) mögliche äquivalente Aufgabe ist die Lösung der Variationsungleichung in H

$$(6) \qquad (g - Tu,u - v) \geq 0 \, , \quad v \in C \quad \text{(vgl. z. B. [6]).}$$

4. Variationsprobleme mit Nebenbedingungen

Wir betrachten in den Voraussetzung von Punkt 2

(7) $$f(u) = \min_{v \in C \subset H} f(v) .$$

Hier ist die Äquivalenz zu (6) mit $g = 0$ und $T = f'$ möglich [11, 12].

5. Gewöhnliche Differentialgleichungen im Hilbert-Raum

Gesucht ist die Lösung des Anfangswertproblems $x(0) = d \in H_1$ über dem Intervall $[0,T]$, $T > 0$, für die abstrakte Differentialgleichung in H_1

(8) $$x'(t) = J(t,x(t))$$

mit der Bochner-integrablen rechten Seite $J(t,x(t))$. Für in x lipschitzstetige J finden wir die Äquivalenz zu einem Fixpunktproblem $x = A^m x$ mit dem strikt kontraktiven Operator A^m : $Ax(t) \equiv d + \int_0^t J(s,x(s))\, ds$ [8].

6. Optimale Steuerung mit abstrakten Differentialgleichungen

Gesucht ist ein $u_0 \in C \subset L_2(0,T;H_0) \equiv H$, daß

(9) $$f(u_0) = \min_{u \in C} f(u)$$

mit $f(u) = \int_0^T F(t,x(t),u(t))\, dt + G(x(T))$, wobei die "Phasenkoordinate" $x(t)$ und der "Steuerparameter" $u(t)$ durch die den Prozeß beschreibende lineare abstrakte Differentialgleichung

$$x'(t) = A(t)x(t) + B(t)u(t) , \quad x(0) = d \in H_1 ,$$

miteinander verbunden sind [9].

7. Nichtlineare Eigenwertprobleme

In engem Zusammenhang mit Minimumproblemen der Form

$$f(u) = \min f(v) : v \in \{x \in H : h(x) = \min h(y) , y \in H\}$$

stehen nichtlineare Eigenwertprobleme der Form

(10) $$f'(u) = zh'(u) , \quad z \in R_1 .$$

Die in den Aufgaben 1 - 7 auftretenden Grundtypen von abstrakten Problemen im Hilbert-Raum H lassen sich durch zwei nichtlineare

Probleme in H mit der konvexen abgeschlossenen Vergleichsmenge $C \subset H$ erfassen:

I. Minimumprobleme mit Nebenbedingungen. $f(u) = \min_{v \in C} f(v)$.

II. Nichtlineare Variationsungleichungen. $(g - Tu, u - v) \geq 0$, $v \in C$.

Für $g = 0$, konvexe f und hemistetige $T = f'$ sind I und II äquivalent. Weiter erfaßt II speziell Operatorgleichungen (1) und Fixpunktprobleme (3). II läßt sich auch allgemeiner fassen:

III. $(Au - u, u - v) \geq r(u) - r(v)$, $v \in C$, mit $r : H \to R_1$.

III erfaßt I auch für nichtdifferenzierbare f . Wenn wir für $u \in H$ und unterhalbstetiges konvexes r die einzige Lösung des Problems $\min_{v \in C} \{\frac{1}{2}\|u - v\|^2 + r(v)\}$ mit $P(C,r)u$ bezeichnen, dann ist III äquivalent dem Fixpunktproblem

IV. $u = P(C,r)Au$.

Mit Fragen der Existenz von Lösungen der Aufgabe III in speziellen Fällen haben sich vor allem F. E. Browder [18, 19], H. Brezis [20] und J. L. Lions [21, 22] auseinandergesetzt. Wir weisen auch auf die in [11, 12, 17] über IV geführten Existenzbeweise hin.

Wir führen nun kurz einige abstrakte Varianten für die Nebenbedingungen C an:

a) $C = \{x \in H : h(x) = \min h(y) , y \in H\}$ für $h : H \to R_1$.

b) $C = \{x \in H : Sx = 0\}$ für $S \in (H \to H)$.

c) $C = \{x \in H : h(x) \leq 0\}$.

d) $C = \{x \in H : Bx = x\}$ für $B \in (H \to H)$.

e) $C = \{x \in H : x$ ist eine Lösung der Variationsungleichung $(-Wz, z - w) \geq h(z) - h(w)$ für $w \in K \subset H\}$ mit $W \in (H \to H)$.

B. NUMERISCHE VERFAHREN

Wir führen hier numerische Verfahren zur approximativen Lösung von I und II an. In den ausgewiesenen Veröffentlichungen finden sich die Verfahren zum großen Teil auch für die allgemeinere Aufgabe III.

Neben den vier Grundtypen

(i) Iterationsverfahren,
(ii) Projektionsverfahren (speziell Ritz- und Galerkin-Verfahren),
(iii) Methoden der Straffunktionale und der Strafoperatoren,
(iv) Elliptische und Tichonoffsche Regularisierung

spielen konstruktivere numerische Verfahren in der Form von Kombinationen dieser Grundtypen eine Rolle. Sie enthalten als Spezialfälle die Verfahren (i) - (iv).

Die Grundtypen (i) - (iv) und die Kombinationen (ii) + (iii) und (ii) + (iv) lassen sich als Folgen von nichtlinearen Variationsungleichungen bzw. als Folgen von Fixpunktproblemen IV fassen, und die Kombinationen der eben genannten Verfahren mit Iterationsvorschriften

$$z_i = T_i z_{i-1}$$

stellen Iterationsverfahren bei Folgen von nichtlinearen Variationsungleichungen III bzw. bei Folgen von Fixpunktproblemen IV $(u_i = T_i u_i)$ dar.

Die betreffenden Veröffentlichungen enthalten neben den Konvergenzuntersuchungen für die hier anzuführenden und für weitere numerische Verfahren Aussagen über Konvergenzgeschwindigkeit, Stabilität bezüglich Fehler bei der numerischen Realisierung der Verfahren sowie Fehlerabschätzungen.

Im weiteren sei (H_i) eine monotone Folge von Unterräumen H_i von H, $H_i \subset H_j$ für $i \leq j$, P_i der Orthogonalprojektor von H auf H_i (Es gilt $P_i = P(H_i, 0)$.) und $\|P_i u - u\| \to 0$ für $i \to \infty$ und $u \in H$.

Wir gehen nun auf einige der genannten numerischen Verfahren ein.

1. Die Probleme (1) und (3) sind z. B. für lipschitzstetige stark monotone T bzw. für strikt kontraktive A über das Galerkin-Verfahren $u_i = P_i A u_i$ $(u_0 = A u_0)$ lösbar. Dabei kann speziell für das verallgemeinerte Dirichletsche Problem im Fall $u_0 \in C^s(G)$, $s > \ell$, unter entsprechenden weiteren Voraussetzungen folgende Konvergenzgeschwindigkeit angegeben werden:

$$\|u_0 - u_i\| \leq M_0 / i^{s-\ell} \;; \quad M_0 \text{ ist fixiert } [4, 11] .$$

Für nur monotone lipschitzstetige T läßt sich bei Existenz von Lösungen der Gleichung (1) bzw. des Problems (3) die zu $v_0 \in H$ nächste Lösung über ein regularisiertes Iterationsverfahren

$$x_i = (W_i)^{n(i)} x_{i-1} \qquad [2 - 4, 11]$$

mit $W_i = I - t_i T_{a_i}$, $T_{a_i} = (T - g) + a_i(I - v_0)$, $a_i \to +0$, über ein regularisiertes Projektionsverfahren

$$P_i T_{a_i} u = u \qquad [2 - 4, 11]$$

und ein regularisiertes Projektions-Iterationsverfahren

$$x_i = (P_i W_i)^{n(i)} x_{i-1} \qquad [2 - 4, 11]$$

näherungsweise bestimmen. Zu Fragen der Existenz von Lösungen für (1) vgl. z. B. [18] und für (3) z. B. [2].

Unter Berücksichtigung des Zusammenhanges von (1) und (3) zu (4) sind die Aussagen auf (4) anwendbar [11, 12]. Daneben läßt sich für stark konvexe stetige f mit hemistetigen f' das Projektions-Iterationsverfahren

$$x_{n+1} = P_n\left(x_n - \frac{f'(x_n)}{n\|P_n f'(x_n)\|}\right), \quad x_1 \in H_1 ,$$

anwenden [5].

2. Wir betrachten nun III in der Form

IIIa. $(Au - u, u - v) \geqq 0$, $v \in C$.

Dabei sei $C_i = C \cap H_i$ und $C_1 \neq \emptyset$. Für strikt kontraktive A ist hier die einzige Lösung über das Projektions-Iterationsverfahren

$$x_i = P(C_i, 0) P_i A x_{i-1} \qquad [11, 12]$$

für den Fall $C = \{x \in H : h(x) = \min h(y) , y \in H\}$ über die Verknüpfung (ii) + (iii) in der Form

$$(Au - u, u - v) \geqq [h(u) - h(v)]/a_i , \quad v \in H_i , \qquad [11, 12]$$

die entsprechende iterierte Methode

$$x_i = P(H_i, h/a_i) P_i A x_{i-1} , \qquad [11, 12]$$

für $C = \{x \in H : Bx = x\}$ bzw. $C = \{x \in H : Sx = 0\}$ über eine weitere Verknüpfung von (ii) und (iii)

$$P_i[k_i Bu + (1 - k_i)Au - u] = 0 , \quad u \in H_i , \qquad [3, 11, 12]$$

und die entsprechend iterierte Methode

$$x_i = (P_i W_i)^{n(i)} x_{i-1} , \quad 0 < k_i < 1 , \quad k_i \to 1 ,$$

$W_i = (1 - t)I + t(k_i B + (1 - k_i)A)$, für bestimmte $t \in R_1$ und $n(i)$ approximierbar [3, 16]. Daneben ist für II (bzw. für I mit $f' = T$ und $g = 0$) bei stark monotonen hemistetigen Potentialoperatoren T im Fall $C = \{x \in H : h(x) \leq 0\}$ das Projektions-Iterationsverfahren

$$x_{n+1} = P_n\left(x_n - \frac{(T - g)x_n}{n\|P_n(T - g)x_n\|}\right), \quad x_n \in C_n ,$$

$$= P_n\left(x_n - \frac{h'(x_n)}{\|nP_n(h'(x_n))\|}\right), \quad x_n \notin C_n$$

anwendbar [6].

Für mehrdeutig lösbare IIIa bzw. II mit lipschitzstetigen pseudokontraktiven A (bzw. lipschitzstetigen monotonen $T = I - A$) ist die zu $v_0 \in H$ nächste Lösung über das regularisierte Iterationsverfahren

$$x_i = (P(C,0)W_i)^{n(i)} x_{i-1} , \qquad [16]$$

$W_i = I - t(I - A_i)$, $A_i = k_i A + (1 - k_i)v_0$, und das regularisierte Projektions-Iterationsverfahren

$$x_i = (P(C_i,0)P_i W_i)^{n(i)} x_{i-1} \qquad [11, 12]$$

für bestimmte t und $n(i)$ näherungsweise lösbar.

3. Eine Reihe der hier angeführten numerischen Verfahren reduziert sich in jedem Schritt auf die Lösung linearer algebraischer Gleichungssysteme, wenn von der Realisierung der auftretenden Skalarprodukte abgesehen wird.

4. Die Arbeiten [5, 11] enthalten Anwendungen der hier angeführten Ergebnisse auf das verallgemeinerte Dirichletsche Problem, [11] Anwendungen auf Probleme der nichtlinearen Kontinuumsmechanik, [11, 6] auf Variationsungleichungen für Differentialoperatoren in partiellen Ableitungen, [8, 10] auf abstrakte Differentialgleichungen und [9] auf Probleme der optimalen Steuerung.

L I T E R A T U R

[1] R. Kluge. Mber. Dt. Akad. Wiss. 11 (1969), 599 - 609.

[2] R. Kluge. Mber. Dt. Akad. Wiss. 12 (1970), 85 - 97.

[3] R. Kluge. Mber. Dt. Akad. Wiss. 12 (1970), 120 - 134.

[4] R. Kluge. Mber. Dt. Akad. Wiss. 12 (1970), 237 - 249.

[5] R. Kluge. Mber. Dt. Akad. Wiss. 12 (1970), 721 - 734.

[6] R. Kluge. Math. Nachr. 48 (1971), 341 - 352.

[7] R. Kluge. Math. Nachr. 51 (1971), 343 - 356.

[8] R. Kluge: Zur Existenz und numerischen Approximation von Lösungen abstrakter Differentialgleichungen erster Ordnung mit Bochner-integrablen Seiten. Math. Nachr., im Druck.

[9] R. Kluge: Näherungsverfahren für ein Problem der optimalen Steuerung mit abstrakten Differentialgleichungen erster Ordnung. Math. Nachr., im Druck.

[10] R. Kluge: Über einige Klassen abstrakter Differentialgleichungen beliebiger Ordnung im Banach-Raum. Mber. Dt. Akad. Wiss., im Druck.

[11] R. Kluge: Dissertation B. Berlin 1970.

[12] R. Kluge: Folgen und Iterationsverfahren bei Folgen nichtlinearer Variationsungleichungen mit strikt kontraktiven Operatoren. Manuskript.

[13] R. Kluge: Numerische Verfahren für nichtlineare analytische Probleme im eindeutig wie im mehrdeutig lösbaren Fall. Sammelband. Gründung und Eröffnung des Weiterbildungszentrums Mathematische Kybernetik und Rechentechnik der TU Dresden, im Druck.

[14] H. Gajewski, R. Kluge. Mber. Dt. Akad. Wiss. 12 (1970), 98 - 115.

[15] H. Gajewski, R. Kluge. Mber. Dt. Akad. Wiss. 12 (1970), 145 - 149.

[16] H. Gajewski, R. Kluge. Mber. Dt. Akad. Wiss. 12 (1970), 405 - 418.

[17] H. Gajewski, R. Kluge. Math. Nachr. 46 (1970), 363 - 373.

[18] F. E. Browder: Problèmes non-linéaires. Montreal Press 1966.

[19] F. E. Browder: Existence and approximation of solutions of nonlinear variational inequalities. Proc. Nat. Acad. Scien. 56 (1966), 1080 - 1086.

[20] H. Brezis: Équations et inéquations non-linéaires dans les espaces vectoriels en dualité. Ann. Inst. Fourier Grenoble 18 (1968), 115 - 175.

[21] J. L. Lions: Über Ungleichungen in partiellen Ableitungen. Uspechi mat. nauk XXVI, 2 (1971), 205 - 263.

[22] J. L. Lions: Quelques méthodes de résolution des problèmes aux limites non linéaires. Dunod, Gauthier - Villars, 1969.

Zentralinstitut für Mathematik und Mechanik der Deutschen Akademie
der Wissenschaften zu Berlin, Mohrenstr. 39, 108 Berlin, DDR

ZUR NUMERISCHEN APPROXIMATION VON NICHTLINEAREN EIGENWERTPROBLEMEN UND FIXPUNKTBIFURKATIONEN FÜR MEHRDEUTIGE ABBILDUNGEN

REINHARD KLUGE, BERLIN (DDR)

I. ZUR NUMERISCHEN APPROXIMATION VON NICHTLINEAREN EIGENWERTPROBLEMEN

Sei H ein reeller Hilbert-Raum mit dem Skalarprodukt $(\cdot,\cdot)$, der Norm $\|\cdot\|$, dem Nullelement o und der identischen Abbildung I, $(H \to H)$ die Menge aller Abbildungen von ganz H in H und $T, A, D \in (H \to H)$. $\rightharpoonup$ kennzeichne die schwache und $\to$ die starke Konvergenz in H.

Wir untersuchen hier mittels funktionalanalytischer Methoden folgende Probleme auf Existenz von Lösungen und ihre approximative Bestimmung über Projektions- und Projektions-Iterationsverfahren:

Probleme mit koerzitiven Abbildungen:

(A) Gesucht ist ein $u \in H$ mit

$$Tu = Du + g, \quad g \in H. \tag{1}$$

(A_1) Gesucht ist eine Lösung von (1) $u \in H$ mit $u \neq o$ für $g = o$ (Eigenwertproblem).

Probleme mit nichtkoerzitiven Abbildungen:

(B) Sei $A = I - T$, $C \subset H$ nicht leer und $Ax + Dy \in C$ für $x, y \in C$. Gesucht ist ein Fixpunkt u von $A + D$ in C, d. h. ein $u \in C$ mit

$$u = Au + Du. \tag{2}$$

(B_1) Gesucht ist eine Lösung $u \in C$ von (2) mit $u \neq o$ (Eigenwertproblem).

I.1. Koerzitive Aufgaben

Seien T und D hemistetige Potentialoperatoren, T monoton und D kompakt. Nach Poljak-Berger ist $T - D - g$ der Gradient des Funktionals

$$F(u) = \int_0^1 (T(su),u)\,ds - \int_0^1 (D(su),u)\,ds - (g,u) .$$

1.0.

Satz 1. [2] Wenn ein $R > 0$ derart existiert, daß $F(u) > F(o)$ für $\|u\| > R$, dann besitzt (1) für jedes $g \in H$ wenigstens eine Lösung.

Bemerkung. Für $F(u) \to +\infty$ falls $\|u\| \to \infty$ ist die Voraussetzung von Satz 1 erfüllt.

Satz 2. [2] In den Voraussetzungen von Satz 1 sei $To = Do = o$ und $G \equiv \{u \in H : F(u) < 0\}$ nicht leer. Dann besitzt (A_1) wenigstens eine Lösung.

$G \neq \emptyset$ wird garantiert z. B. durch

Lemma 1. [2] Die hemistetigen Potentialoperatoren T und D genügen den folgenden Voraussetzungen:

a) $To = Do = o$.

b) $(Tx,x) \leq c\|x\|^2 + r(\|x\|)$ mit $c > 0$ und $r(t)/t^2 \to 0$ mit $t \to +0$ für die stetige reellwertige Funktion $r(t)$, $t \geq 0$.

c) Es existiert ein homogener Operator D_0 , daß $(Dx,x) \geq$ $\geq (D_0x,x) - w(\|x\|)$ mit $w(t)/t^2 \to 0$ für $t \to +0$ für die stetige reellwertige Funktion $w(t)$, $t \geq 0$.

d) Es gilt $\sup_{x \in H} ((D_0x,x)/\|x\|^2) = c_0 < +\infty$ und $c_0 > c$.

Dann ist G nicht leer.

Beispiel. Sei T ein lipschitzstetiger Potentialoperator mit $To = o$ und D_0 ein linearer beschränkter selbstadjungierter nichtnegativer Operator. Wir betrachten das Problem

$$Tu = zD_0u , \quad z \in R_1 .$$

Nach Lemma 1 ist für jedes $z > L/\|D_0\|$ mit der Lipschitzkonstanten L von T $G(z)$ nicht leer [3].

1.1. Sei H separabel, (w_i) ein vollständiges System linear unabhängiger Elemente, H_n die lineare Hülle der ersten i_n Elemente von (w_i) und P_n der Projektor von H auf H_n.

Wir sagen, daß $A \in (H \to H)$ der Bedingung $(S)_0$ genügt, wenn aus $u_n \rightharpoonup u$, $Au_n \rightharpoonup z$ und $(Au_n, u_n) \to (z,u)$ folgt $u_n \to u$.

Satz 1.1. [2] In den Voraussetzungen von Satz 1 existieren Ritzsche Näherungsfolgen, d. h. Folgen (u_n) von Lösungen u_n der Minimumprobleme

$$F(u_n) = \min_{v \in H_n} F(v) , \tag{3}$$

die Galerkin-Näherungen sind, d. h. Lösungen der Probleme

$$(Tu - Du - g, h) = 0 , \quad h \in H_n . \tag{4}$$

Wenn weiter T beschränkt ist und $(S)_0$ genügt und D auf $V(o,R) \equiv \{x \in H : \|x\| \leq R\}$ gleichmäßig stetig ist, dann konvergiert jede schwach konvergente Teilfolge der (u_n) auch stark gegen dieselbe Lösung. Ist die Lösung eindeutig bestimmt, dann konvergiert die gesamte Folge (u_n) gegen diese Lösung.

Satz 2.1. [2] In den Voraussetzungen von Satz 1.1 sei $To = Do = o$ und G nicht leer. Dann konvergieren die Folgen der Lösungen von (3) und (4) wie in Satz 1.1 gegen Eigenfunktionen, d. h. gegen Lösungen von (A_1).

1.2. Sei $S_n = P_n S P_n$ für $S \in (H \to H)$ und $g_n = P_n g$.

Satz 1.2. [1] In den Voraussetzungen von Satz 1 seien T und D lipschitzstetig mit den Konstanten L_T und L_D und $T - D$ koerzitiv, $((T - D)u,u)/\|u\| \to +\infty$ für $\|u\| \to \infty$. Dann ist das Projektions-Iterationsverfahren

$$u_n = P_n u_{n-1} - aT_n u_{n-1} + aD_n u_{n-1} + ag_n \tag{5}$$

für $0 < a < 2/(L_T + 3L_D)$ beschränkt. Wenn dabei T $(S)_0$ genügt, dann konvergiert jede schwach konvergente Teilfolge von (5) auch stark gegen die gleiche Lösung von (1). Existiert nur eine Lösung, dann konvergiert die gesamte Folge (5) in der Norm gegen diese.

Satz 2.2. [2] In den Voraussetzungen von Satz 1.2 sei $To = Do = o$, $g = o$, G nicht leer und $u_0 \in G$. Dann konvergiert jede schwach konvergente Teilfolge von (5) auch stark gegen die gleiche Eigenfunktion.

Bemerkung. (5) reduziert sich neben der Berechnung von Skalarprodukten in jedem Schritt auf die Lösung eines endlichen linearen algebraischen Gleichungssystems

$$(u_n, w_r) = (u_{n-1} - aTu_{n-1} + aDu_{n-1} + ag, w_r) , \quad r = 1, \ldots, i_n ,$$

bezüglich der Koeffizienten der Zerlegung von u_n nach den Koordinatenelementen w_r, $r = 1, \ldots, i_n$.

I.2. Nichtkoerzitive Aufgaben

Satz 3. [4, 5] Sei $I - A$ monoton, A lipschitzstetig, D verstärkt stetig ($Dx_n \to Dx$ für $x_n \rightharpoonup x$) und C abgeschlossen, konvex und beschränkt. Dann besitzt B wenigstens eine Lösung ($A + D$ wenigstens einen Fixpunkt in C.)

Es seien nun die Voraussetzungen von Punkt 1.1 erfüllt.

Satz 3.1. [4, 5] Sei in den Voraussetzungen von Satz 3 $C \equiv \{x \in H : \|x\| \leq R\}$. Dann existieren Galerkin-Folgen (u_n), d. h. Folgen von Lösungen u_n der Probleme

$$(u - Au - Du, h) = 0 , \quad h \in H_n ,$$

die in C enthalten sind. Wenn $(I - A)$ $(S)_0$ genügt, dann konvergiert jede schwach konvergente Teilfolge von (u_n) auch stark in H gegen den gleichen Fixpunkt.

Satz 3.2. [4, 5] A, D seien Potentialoperatoren, $Ax + Dx \in C$ für $x \in C \equiv \{y \in H : \|y\| \leq R\}$, $I - A$ monoton und stetig, D verstärkt stetig, A und D auf C lipschitzstetig und $u_0 \in C$. Dann ist das Projektions-Iterationsverfahren

$$(6) \qquad u_n = (1 - a)P_n u_{n-1} + aA_n u_{n-1} + aD_n u_{n-1}$$

für $0 < a < \min\{1, 2/(1 + L_A + 3L_D)\}$ in C enthalten. $A + D$ besitzt wenigstens einen Fixpunkt in C. Wenn $(I - A)$ $(S)_0$ genügt, dann konvergiert jede schwach konvergente Teilfolge auch stark gegen den gleichen Fixpunkt. Bei Existenz genau einer Lösung konvergiert (6) ganz gegen diese.

Satz 4.2. In den Voraussetzungen von Satz 3.2 sei entweder $o \notin C$ oder $Ao = Do = o$, $G(C) \equiv \{x \in C : F(x) < 0$; $T = I - A$,

$g = o\}$ nicht leer und $u_0 \in G(C)$. Dann besitzt (B_1) wenigstens eine Lösung und (6) konvergiert wie in Satz 3.2 gegen entsprechende Eigenfunktionen.

II. FIXPUNKTBIFURKATION FÜR MEHRDEUTIGE ABBILDUNGEN

Wir betrachten das nichtlineare Eigenwertproblem

$$Tu = D(z)u \ , \quad z \in M \neq \emptyset \ . \tag{7}$$

Existiert die eindeutige Abbildung $T^{-1} \in (H \to H)$, dann ist (7) äquivalent zu

$$u = A(z)u \quad \text{mit} \quad A(z) = T^{-1}D(z) \ . \tag{8}$$

Ist T^{-1} mehrdeutig, dann ist (7) äquivalent zu

$$u \in A(z)u \ . \tag{9}$$

Wir geben hier Ergänzungen zur Leray-Schauderschen Theorie für mehrdeutige Abbildungen und Verallgemeinerungen für die Sätze von M. A. Krasnoselskij über die Bifurkation von Eigenfunktionen aus der Nullösung für eindeutige Abbildungen auf mehrdeutige, u. z. für eine spezielle Parameterabhängigkeit. Auf die Realisierung der hier angedeuteten Methode der mehrdeutigen Regularisatoren gehen wir an einer anderen Stelle ein.

Sei B ein reeller Banach-Raum und $H(B)$ die mit der Hausdorff-Metrik versehene Menge aller nichtleeren, abgeschlossenen und beschränkten Teilmengen von B.

Definition. Die mehrdeutige Abbildung A von B in B besitzt in $o \in B$ die Fréchetsche Ableitung C, wenn $Ax - Ao = Cx + Wx$, $x \in B$, mit der linearen oberhalbstetigen Abbildung C und $\sup_{y \in Wx} \|y\|/\|x\| \to 0$ für $\|x\| \to 0$.

Jede mehrdeutige Abbildung A mit $Ax \in H(B)$ erzeugt eine eindeutige Abbildung $\overline{A} : B \to H(B)$ $(\overline{A}x = \{Ax\})$.

Satz 5. [9] Sei o Fixpunkt der kompakten Abbildung $A \in (B \to B)$ mit $Ao = \{o\}$, für die Ax, $x \in B$, konvex und abgeschlossen und $\overline{A}$ stetig ist. Sei weiter 1 kein Eigenwert der eindeutigen

Fréchetschen Ableitung C von A in o . Dann ist o ein isolierter Fixpunkt von A und sein Index gleich $(-1)^b$, wobei b die Summe der (algebraischen) Vielfachheiten der in (0,1) gelegenen charakteristischen Zahlen von C ist.

Es sei Ao = {o} . Weiter besitze A in o die eindeutige Fréchetsche Ableitung C . Wir untersuchen das Problem

$$u \in zAu \ , \quad z \in R_1 \ ,$$

auf Bifurkation aus dem Nullelement.

Satz 6. [9, 10] 1. Sei C kompakt. Dann ist jeder Bifurkationspunkt von A eine charakteristische Zahl von C .

2. Sei A kompakt, $\bar{A}$ stetig und Ax konvex und abgeschlossen für alle $x \in B$. Dann ist jede charakteristische Zahl von C mit ungerader (algebraischer) Vielfachheit ein Bifurkationspunkt von A .

Bemerkungen. 1. Für eindeutige Abbildungen A resultieren aus Satz 6 sofort die bekannten Sätze von M. A. Krasnoselskij [11].

2. Bezüglich der Bifurkationstheorie mit allgemeinen metrischen Parametern vgl. für eindeutige Abbildungen [6, 7] und für mehrdeutige Abbildungen [8, 9].

L I T E R A T U R V E R Z E I C H N I S

[1] R. Kluge: Ein Projektions-Iterationsverfahren bei Fixpunktproblemen und Gleichungen mit monotonen Operatoren. Mber. Dt. Akad. Wiss. 11 (1969), 599 - 609.

[2] R. Kluge: Zur numerischen Approximation von Eigenfunktionen in nichtlinearen Eigenwertproblemen. Mber. Dt. Akad. Wiss. 11 (1969), 811 - 818.

[3] R. Kluge: Zur Existenz und Approximation von Eigenfunktionen in allgemeinen nichtlinearen Eigenwertproblemen. G. Anger. Elliptische Differentialgleichungen, Bd. I, 57 - 60 (1970).

[4] R. Kluge: Zur approximativen Lösung nichtlinearer Variationsungleichungen. Dissertation B. Berlin 1970.

[5] R. Kluge: Über eine Klasse gestörter Fixpunktprobleme. Math. Nachr. 47 (1970), 319 - 329.

[6] R. Kluge: Zur Existenz und Realisierungsweise von Bifurkationselementen. Math. Nachr. 42 (1969), 173 - 192.

[7] R. Kluge: Zur Lösung eines Bifurkationsproblems für die Karmanschen Gleichungen im Fall der rechteckigen Platte. Math. Nachr. 44 (1970), 29 - 54.

[8] R. Kluge: Fixpunktbifurkation für parameterabhängige vieldeutige vollstetige Abbildungen. Mber. Dt. Akad. Wiss. 11 (1969), 89 - 95.

[9] R. Kluge: Differenzierbare mehrdeutige Abbildungen und Fixpunktbifurkation. Mber. Dt. Akad. Wiss. 11 (1969), 469 - 480.

[10] R. Kluge: Fixpunktbifurkation mehrdeutiger Abbildungen. G. Anger. Elliptische Differentialgleichungen, Bd. I, 53 - 55 (1970).

[11] M. A. Krasnoselskij: Topologische Methoden in der Theorie der nichtlinearen Integralgleichungen (russ.). Moskau 1956.

Zentralinstitut für Mathematik und Mechanik der Deutschen Akademie der Wissenschaften zu Berlin, Mohrenstr. 39, 108 Berlin, DDR

MORSE-SARD THEOREM FOR FUNCTIONS FROM THE CLASS $C^{k,\alpha}$

MILAN KUČERA, PRAHA (CZECHOSLOVAKIA)

This paper deals with the critical-value problem of real functions. Let us explain what led us to this problem. One of the main problems of nonlinear spectral analysis is how many eigenvalues nonlinear operators can have. Let us consider two operators T, S which map a Banach space X into its dual X^*. If T, S are potential operators which have some suitable properties, then the following assertion is true:

If λ is an eigenvalue of T, S and x_0 is the corresponding eigenvector, i. e. $\|x_0\| \neq 0$, $\lambda T x_0 - S x_0 = 0$, then $\lambda = \varphi(x_0)$ and x_0 is a critical point of the functional $\varphi(x) = (Sx,x)/(Tx,x)$.

Hence, the problem of eigenvalues leads to the following problem of critical values: How big can be a set of all critical values of a real functional which is defined on a Banach space X. We say that γ is a critical value of the functional φ if $\gamma = \varphi(x)$, where x is a critical point of φ, i. e. the derivative of φ (in a certain sense) is equal to zero at x.

In this paper the full answer to this question is given in the case $X = E_n$, where E_n is the n-dimensional Euclidean space. All results formulated here are proved in [3].

Let us consider a function f defined on an open set Ω in E_n. Denote $Z_f = \{x \in \Omega;\ \frac{\partial f}{\partial x_i}(x) = 0,\ i = 1, 2, \ldots, n\}$. The following Morse-Sard Theorem is well-known:

If $f \in C^n(\Omega)$, then the set of critical values $f(Z_f)$ has Lebesgue measure equal to zero.

This theorem can be generalized and made more precise (see [2], [5], [6]).

We shall write $f \in C^{k,\alpha}(\Omega)$ (for a positive integer k and

$\alpha \in \langle 0,1 \rangle$), if the function f has continuous partial derivatives in Ω of all orders not exceeding k and if all derivatives of the order k are α-Hölderian.

Let M be a subset of E_1. For each positive number s we define s-Hausdorff measure $\mu_s(M)$ of the set M in the following way: for each $\varepsilon > 0$ we put

$$\mu_{s,\varepsilon}(M) = \inf \sum_{i=1}^{\infty} (\operatorname{diam} A_i)^s ,$$

the infimum being taken over all countable coverings $\{A_i\}_{i=1}^{\infty}$ of M such that $\operatorname{diam} A_i < \varepsilon$, $i = 1, 2, \ldots$; we define $\mu_s(M) = \lim_{\varepsilon \to 0+} \mu_{s,\varepsilon}(M)$. If $\mu_s(M) = 0$, then we say M is s-null. It is easy to see that if M is s-null, then M is r-null for each $r > s$. If $s = 1$, then we obtain Lebesgue measure.

Main Theorem. If $f \in C^{k,\alpha}(\Omega)$, then $f(Z_f)$ is $\frac{n}{k+\alpha}$-null. No better result holds, because if $s < \frac{n}{k+\alpha}$, then a function $f \in C^{k,\alpha}$ can be constructed such that $\mu_s(f(Z_f)) > 0$.

Special cases of the theorem are proved in [1] (for $n = 1$) and in [2], [5], [6] (for arbitrary n and $\alpha = 0$).

Remark 1. If $f \in C^{\infty}(\Omega)$, then the set $f(Z_f)$ is s-null for each s positive. But this set need not be countable. We must demand f real-analytic to obtain such a strong result (see [4]).

Proof of Main Theorem consists of two parts, namely of the proofs of the following two theorems:

Theorem 1. Suppose $f \in C^1(\Omega)$ and r is a real number, $r \geq 1$. Let A be a closed subset of Z_f. Assume

$$|f(x') - f(x)| \leq C\|x' - x\|^r \tag{1}$$

for each $x', x \in A$ (where $C > 0$).

Then the set $f(A)$ is $\frac{n}{r}$-null.

Theorem 2. If $f \in C^{k,\alpha}(\Omega)$, then there exists a countable system $\{M_t\}_{t=1}^{\infty}$ of sets such that

(a) $Z_f \setminus \bigcup_{t=1}^{\infty} M_t$ is countable;

(b) for each positive integer t there exists $C > 0$ (depending on t) such that $|f(x') - f(x)| \leq C\|x' - x\|^{k+\alpha}$ for each x', $x \in M_t$.

The basic idea of the proof of Theorem 1 is taken from the paper [1]. This theorem being proved, our Main Theorem is clear for $n = 1$. Let us consider this case. It is sufficient to prove that $f(Z_f')$ is $\frac{1}{k+\alpha}$-null, where Z_f' is the set of all limit points of Z_f (because the set $f(Z_f) \setminus f(Z_f')$ is countable). If $x \in Z_f'$, then all derivatives of f at x which exist are equal to zero. We obtain $|f(x') - f(x)| \leq C\|x' - x\|^{k+\alpha}$ by using Taylor's Theorem. It follows from Theorem 1 that $f(Z_f')$ is $\frac{1}{k+\alpha}$-null.

In the general case $n > 1$ we cannot use this argument, because the derivatives of higher orders at a limit point of the set Z_f can be different from zero. We must use Theorem 2.

Theorem 2 generalizes some result of [2], where the case $\alpha = 0$ (i. e. $f \in C^k(\Omega)$) is considered. The assertion of Theorem 2 is clear in the one-dimensional case as we have seen before. A. P. Morse proved this theorem (with $\alpha = 0$) by using induction over $m = n + k$. Theorem 2 can be proved in a similar way. However, in [3] a constructive proof is given. This proof is based on reducing the dimension of the domain by using the implicit function theorem and on the fact that each set M_t lies in some hyperplane.

Remark 2. A similar result as Main Theorem can be proved for Fredholm functionals on infinite dimensional Banach spaces. Application of such results are given in the paper [7].

REFERENCES

[1] R. Kaufmann: Representation of linear sets as critical sets. Proc. Amer. Math. Soc. (25) 1970, No. 4.

[2] A. P. Morse: The behaviour of a function on its critical set. An. Math. (2) 40 (1939).

[3] M. Kučera: Hausdorff measures of the set of critical values of functions of the class $C^{k,\lambda}$. Comm. Math. Univ. Car. 13, 2 (1972).

[4] J. Souček, V. Souček: The Morse-Sard theorem for real-analytic functions. Comm. Math. Univ. Car. 13, 1 (1972).

[5] A. Sard: The measure of the critical set values of differentiable mappings. Bull. Amer. Math. Soc. 48 (1942).

[6] A. J. Dubovickij: Sets of points of degeneration. Izvestija AN SSSR 31 (1967).

[7] S. Fučík, M. Kučera, J. Nečas, J. Souček, V. Souček: Morse-Sard theorem in infinite dimensional Banach spaces and investigation of the set of all critical levels. To appear.

Matematický ústav ČSAV, Žitná 25, Praha 1, Czechoslovakia

REMARKS ON NONLINEAR EIGENVALUE PROBLEMS

JOACHIM NAUMANN, BERLIN (GDR)

1. INTRODUCTION

The purpose of the present note is to weaken some assumptions of the author's recent paper [10] and to give a remark on the verification of condition (C) due to R. S. Palais and S. Smale.

Our study is based on a careful investigation of an abstract initial value problem in a Banach space X whose solution is a trajectory on the level surface $g^{-1}(R)$ of a functional g on the Banach space X. This trajectory is used in Section 3 to construct a deformation in order to give an indirect proof of the crucial Lemma 3 whose formulation is based on a variational principle (which is due to L. A. Ljusternik and L. G. Schnirelman, see e. g. [9]) for a functional f on the level surface $g^{-1}(R)$. The next section contains two theorems on the existence of a solution of the nonlinear eigenvalue problem $\lambda g'(u) = f'(u)$. The proofs of these theorems employ, besides Lemma 3, either the pseudo-monotonicity or the condition $(S)_0$ for g' and a compactness condition for f'. In the fifth section we give as an application of these results an existence theorem for an infinite sequence of eigenvectors of the above nonlinear eigenvalue problem. The proof is based on the notion "category of a set" and requires the eveness of g and f. We conclude our paper with a remark on the application of the abstract results to nonlinear elliptic eigenvalue problems and a lemma on verification of condition (C) for a functional. This condition plays an important role in the treatment of Ljusternik-Schnirelman theory on differentiable manifolds given by R. S. Palais in [12].

Various generalizations of the variational method of Ljusternik-Schnirelman can be found in the articles by M. S. Berger [1], [2], F.

E. Browder [3], [4], [5], E. S. Citlanadze [6] and in the book by M. M. Vaijnberg [13]. The treatment in [5] contains a verification, by abstract functional analysis techniques, of some properties and conditions for functions which are necessary for the study of this method on Finsler manifolds. Some of our assumptions are weaker than various assumptions which are formulated in this paper. A modification of the mentioned variational method which does not require the eveness of the functionals considered, is given in the book by M. A. Krasnoselskij [8] and in the author´s paper [11].

Recently S. Fučík and J. Nečas ([7]) have given an important new approach to the basic lemma of Ljusternik-Schnirelman theory. The authors have used implicit function theorem technique which makes it possible to dispense with the local Lipschitz condition for the derivative of the functional g, which defines the level surface $g^{-1}(R)$.

2. BASIC LEMMAS

Let X be a real Banach space, X^* its dual space and (w,u) the pairing between $u \in X$ and $w \in X^*$. We denote by $\to$ the strong convergence in X or in X^*, and by $\rightharpoonup$ the weak convergence in X or in X^*.

Denote by g a functional which is defined on X. Analogously to [2], [3], [10] and [11] we introduce for fixed real $R > 0$ the level surface

$$\partial M_R = \partial M_R(g) = \{u \mid u \in X , g(u) = R\} .$$

Let A be a (nonlinear) operator which maps X into X^*. In what follows we suppose particularly that A is the gradient of the functional g considered above. Independently of this, we formulate

Assumption I. The operator A is defined on X, maps X into X^* and satisfies the following conditions:

(i) A satisfies on $\partial M_R(g)$ uniformly the local Lipschitz condition; there exist constants $r = r(R) > 0$ and $k = k(R) > 0$ independent of $v \in \partial M_R(g)$ such that $\|Au_1 - Au_2\| \leq k\|u_1 - u_2\|$ holds for all $u_1, u_2 \in B(v,r) = \{u \mid u \in X , \|u - v\| \leq r\}$ (the constant k does not depend on u_1, u_2).

(ii) For a constant $\alpha = \alpha(R) > 0$ which does not depend on $u \in \partial M_R(g)$ it holds that $|(Av,v)| \geq \alpha$ for all $v \in B(u,r)$ and for all $u \in \partial M_R(g)$ (r taken according to (i)).

(iii) A maps bounded sets into bounded sets.

Suppose ∂M_R is bounded. If $F(v) = h - ((Av,h)/(Av,v))v$ for $v \in \partial M_R$ and arbitrary $h \in X$ with $\|h\| = 1$, it follows by assumption I that

$$\|F(v_1) - F(v_2)\| \leq c_1 \|v_1 - v_2\| \;,\; \|F(v)\| \leq c_2$$

for arbitrary v_1, v_2, $v \in B(u,r)$ uniformly for all $u \in \partial M_R$. The constants c_1 and c_2 in general depend on R ($c_1 = c_1(R)$, $c_2 = c_2(R)$) but not on v_1, v_2, v. With $q \in (0,1)$, we introduce

$$t_0 = \min\left\{\frac{q}{c_1} , \frac{r}{3c_2}\right\} .$$

Obviously the number t_0 depends neither on $v \in B(u,r)$ nor on $u \in \partial M_R$.

We now consider for $h \in X$ with $\|h\| = 1$ the initial value problem

$$\text{(1)} \qquad \frac{du(t)}{dt} = h - \frac{(Au(t),h)}{(Au(t),u(t))} u(t) \;,\; u(0) = u_0 \in \partial M_R$$

and state

<u>Lemma 1</u>. Suppose that A satisfies assumption I and that ∂M_R is bounded.

Then:

(i) There exists in the interval $[0,t_0]$ a unique solution $u(t)$ of (1) with $u(t) \in B(u_0,\frac{r}{3})$.

(ii) For the solution $u(t)$ of (1) it holds that $\|u(t+s) - u(t)\| \leq c_2|s|$ for t, $(t+s) \in [0,t_0]$.

(iii) Moreover, if A is the gradient of the functional g then it holds that $u(t) \in \partial M_R$ for the solution $u(t)$ of (1) ($t \in [0,t_0]$).

<u>Proof of Lemma 1 (i)</u>. We define

$$C(a,b;X) = \text{set of all continuous functions from } [a,b] \text{ into } X .$$

$C(a,b;X)$ is a Banach space with respect to the norm

$$\|u\|_{C(a,b;X)} = \max_{t\in[a,b]} \|u(t)\| .$$

In what follows we identify each point $u \in X$ with the function which maps $t \in [a,b]$ into the constant element $u \in X : u(t) \mapsto u$ for all $t \in [a,b]$.

Let $u_0 \in \partial M_R$ be arbitrarily chosen but fixed. In accordance with our above arrangement we introduce

$$K(u_0,\tfrac{r}{3}) = \{u \mid u \in C(0,t_0;X) \ , \ \|u - u_0\|_{C(0,t_0;X)} \leq \tfrac{r}{3}\} \ .$$

On $K(u_0,\frac{r}{3})$ we define the operator P :

$$u \in K(u_0,\tfrac{r}{3}) \ , \quad P : u \mapsto Pu \ ,$$

$$Pu(t) \equiv u_0 + \int_0^t \left[h - \frac{(Au(s),h)}{(Au(s),u(s))} u(s)\right] ds \ , \quad t \in [0,t_0] \ .$$

Clearly $Pu \in C(0,t_0;X)$.

If $u \in K(u_0,\frac{r}{3})$, it holds

$$\|u(t) - u_0\| \leq \max_{t \in [0,t_0]} \|u(t) - u_0\| \leq \frac{r}{3} \ ,$$

for all $t \in [0,t_0]$, i. e. $u(t) \in B(u_0,\frac{r}{3})$ for all $t \in [0,t_0]$. Hence one obtains

$$\|Pu(t) - u_0\| \leq \int_0^t \left\| h - \frac{(Au(s),h)}{(Au(s),u(s))} u(s) \right\| ds \leq c_2 t \leq \frac{r}{3}$$

and therefore

$$\|Pu - u_0\|_{C(0,t_0;X)} = \max_{t \in [0,t_0]} \|Pu(t) - u_0\| \leq \frac{r}{3} \ ,$$

i. e. $Pu \in K(u_0,\frac{r}{3})$.

For arbitrary u_1, $u_2 \in K(u_0,\frac{r}{3})$ it follows that

$$\begin{aligned} \|Pu_1(t) - Pu_2(t)\| &\leq c_1 \int_0^t \|u_1(s) - u_2(s)\| \, ds \leq \\ &\leq c_1 t \max_{s \in [0,t_0]} \|u_1(s) - u_2(s)\| \leq \\ &\leq q\|u_1 - u_2\|_{C(0,t_0;X)} \end{aligned}$$

for all $t \in [0,t_0]$. Hence

$$\|Pu_1 - Pu_2\|_{C(0,t_0;X)} \leq q\|u_1 - u_2\|_{C(0,t_0;X)} \ .$$

The application of Banach's fixed point theorem concludes the proof, q. e. d.

Proof of Lemma 1 (ii). First, for t , $(t + s) \in [0,t_0]$ one obtains

$$u(t + s) - u(t) = \int_t^{t+s} \left[h - \frac{(Au(\tau),h)}{(Au(\tau),u(\tau))} u(\tau)\right] d\tau$$

and by Lemma 1 (i),

$$\|u(t + s) - u(t)\| \leq c_2 |s| ,$$

q. e. d.

Proof of Lemma 1 (iii). Let $t \in [0,t_0]$ be chosen arbitrarily. For $(t + s) \in [0,t_0]$ we then have $u(t)$, $u(t + s) \in B(u_0,\frac{r}{3})$ for the solution $u(t)$ of (1). Hence it follows $\|(u(t) + \xi[u(t + s) - u(t)]) - u_0\| \leq r$ for each $\xi \in (0,1)$, which implies

$$\|Au(t) - A(u(t) + \xi[u(t + s) - u(t)])\| \leq kc_2|s| .$$

For a suitable mean value $\xi \in (0,1)$ it now holds that

$$\frac{g(u(t + s)) - g(u(t))}{s} =$$

$$= \left(A(u(t) + \xi[u(t + s) - u(t)]), \frac{u(t + s) - u(t)}{s}\right)$$

and the estimation above provides

$$\frac{dg(u(t))}{dt} = \left(Au(t), \frac{du(t)}{dt}\right) = 0 ,$$

i. e. $g(u(t)) \equiv R$ for all $t \in [0,t_0]$, q. e. d.

The solution $u(t)$ of (1) is called a trajectory on ∂M_R .

Let A and B be (nonlinear) operators which map X into X^* . To prepare the considerations of Section 3, let us introduce the mappings Q and $S(.,.)$:

$$Qu = Bu - \frac{(Bu,u)}{(Au,u)} Au , \quad S(v,u) = Bv - \frac{(Bv,u)}{(Au,u)} Au .$$

If A satisfies condition (ii) of assumption I, then obviously Q and $S(v,.)$ are well-defined on each ball $B(u_0,r)$ $(v \in X)$.

Lemma 2. Suppose that A satisfies assumption I and B maps bounded sets into bounded sets. Moreover, let ∂M_R be a bounded set.

Then for arbitrary $u_0 \in \partial M_R$ and arbitrary $u , v \in B(u_0,r)$ we have

$$\|S(v,u) - Qu_0\| \leq c_3(\|Bv - Bu_0\| + \|u - u_0\|) .$$

The constant c_3 $(c_3 = c_3(R))$ depends neither on $u_0 \in \partial M_R$ nor on $u , v \in B(u_0,r)$.

Proof of Lemma 2. First, it holds that

$$\|S(v,u) - Qu_0\| \leq \|Bv - Bu_0\| +$$

$$+ k\left|\frac{(Bv,u)}{(Au,u)}\right| \|u - u_0\| + \left|\frac{(Bv,u)}{(Bu,u)} - \frac{(Bu_0,u_0)}{(Au_0,u_0)}\right| \|Au_0\| .$$

The identity

$$(Au_0,u_0)(Bv,u) - (Au,u)(Bu_0,u_0) =$$

$$= (Bv,u)(Au_0,u_0 - u) + (Bv,u)(Au_0 - Au,u) +$$

$$+ (Bv,u - u_0)(Au,u) + (Bv - Bu_0,u_0)(Au,u) ,$$

easily yields the assertion, q. e. d.

Suppose that the gradient of the functional f defined on X is f' . If $u(t)$ $(t \in [0,t_0])$ is the solution of (1), then it holds for a suitable mean value $\xi = \xi(t) \in (0,1)$:

$$f(u(t)) - f(u_0) = (f'(u_0 + \xi[u(t) - u_0]),u(t) - u_0) .$$

After a simple calculation one obtains

$$f(u(t)) - f(u_0) = \int_0^t (S(\hat{u}_t,u(s)),h)\, ds , \tag{2}$$

$t \in [0,t_0]$, where $\hat{u}_t \equiv u_0 + \xi[u(t) - u_0]$.

3. FORMULATION OF VARIATIONAL PROBLEMS

We now consider ∂M_R as a topological space provided with the topology which is induced by the norm topology of X . Further let

$[V]$ denote the class of subsets $V \subset \partial M_R$ such that $[V]$ contains with a set V each set which is obtained from V by a continuous deformation in ∂M_R.

The definition of the class $[V]$ enables us to formulate for a functional f on X the variational problem

$$(\bullet) \qquad c = c(R) = \sup_{[V]} \inf_{V} f(u) .$$

If f is bounded from below on at least one $V \in [V]$ and if we set $M_a = f^{-1}[a,+\infty)$, it is obvious that similarly to [12],

$$\sup_{[V]} \inf_{V} f(u) = \sup \{a \in R^1 \mid \exists V \in [V] \text{ with } V \subseteq M_a\} .$$

Indeed, if $V \in [V]$ is arbitrarily chosen then $V \subseteq M_a$ with $a = \inf_V f(u)$, i. e.

$$\sup_{[V]} \inf_{V} f(u) \leqq \sup \{a \in R^1 \mid \exists V \in [V] \text{ with } V \subseteq M_a\} ,$$

and by analogous argument we obtain the converse inequality.

We define for $\varepsilon > 0$,

$$W_\varepsilon = \{u \mid u \in \partial M_R , |f(u) - c| < \varepsilon\} .$$

Lemma 3. Let g be a functional which is defined on X, g' being its gradient on X. Suppose that g' satisfies assumption I and that ∂M_R is bounded. Further let f be a functional on X, f' being its gradient on X. Suppose that f' maps bounded sets into bounded sets and that for arbitrary $\eta > 0$ there exists $\delta = \delta(\eta) > 0$ such that $\|f'(u) - f'(u_0)\| \leqq \eta$ holds for all u with $\|u - u_0\| \leqq \delta$ uniformly for all $u_0 \in \partial M_R$.

Then for each $\varepsilon > 0$ there exists (at least) one $u_\varepsilon \in W_\varepsilon$ with $\|Qu_\varepsilon\| < \varepsilon$.

Proof of Lemma 3. Suppose, contrary to our assertion, that there exists an $\varepsilon_0 > 0$ with $\|Qu\| \geqq \varepsilon_0$ for all $u \in W_{\varepsilon_0}$.

Let $u_0 \in W_{\varepsilon_0}$ be chosen arbitrarily. Then there exists $h = h_{u_0} \in X$ with $\|h\| = 1$ and $(Qu_0,h) \geqq \frac{1}{2}\|Qu_0\|$. Further, there exist $\delta_0 > 0$ such that the inequality $\|f'(u) - f'(u_0)\| \leqq \varepsilon_0/(8c_3)$ holds for all u with $\|u - u_0\| \leqq \delta_0$ uniformly with respect to $u_0 \in \partial M_R$ (c_3 taken according to Lemma 2).

We set

$$t_1 = \min \left\{ t_0, \frac{\delta_0}{c_2}, \frac{\varepsilon_0}{8c_2c_3} \right\}$$

(c_2 taken according to Lemma 1 (ii)). By $u(t)$, $t \in [0,t_1]$ we denote the solution of our initial value problem (1)

$$\frac{du(t)}{dt} = h_{u_0} - \frac{(g'(u(t)),h_{u_0})}{(g'(u(t)),u(t))} u(t) , \quad u(0) = u_0 .$$

From Lemma 1 (ii) it follows that

$$\|u(t) - u_0\| \leq c_2 t \leq \delta_0 , \quad t \in [0,t_1]$$

and analogously for the element $\hat{u}_t = u_0 + \xi[u(t) - u_0]$ with $\xi \in (0,1)$,

$$\|\hat{u}_t - u_0\| \leq \|u(t) - u_0\| \leq \frac{\varepsilon_0}{8c_3}, \quad t \in [0,t_1] .$$

Since $\hat{u}_t$, $u(t) \in B(u_0,\frac{r}{3})$, Lemma 2 implies for $t \in [0,t_1]$ the estimate

$$\|S(\hat{u}_t,u(s)) - Qu_0\| \leq \frac{1}{4}\varepsilon_0 \text{ *)}, \quad s \in [0,t_1] .$$

By means of (2) we obtain from this inequality

$$f(u(t)) - f(u_0) \geq \frac{1}{4}\varepsilon_0 t \tag{3}$$

for all $t \in [0,t_1]$ uniformly for all $u_0 \in W_{\varepsilon_0}$.

Let $\varepsilon_1 = \min (\varepsilon_0,\frac{1}{8}\varepsilon_0 t_1)$. For this ε_1 there exists a $V_{\varepsilon_1} \in [V]$ such that $\inf\limits_{V_{\varepsilon_1}} f(u) \geq c - \varepsilon_1$. We set

$$V_{\varepsilon_1}^{(1)} = \{u \mid u \in V_{\varepsilon_1} , c + \varepsilon_1 > f(u) \geq c - \varepsilon_1\} ,$$

$$V_{\varepsilon_1}^{(2)} = \{u \mid u \in V_{\varepsilon_1} , f(u) \geq c + \varepsilon_1\} .$$

Clearly $V_{\varepsilon_1} = V_{\varepsilon_1}^{(1)} \cup V_{\varepsilon_1}^{(2)}$ and $V_{\varepsilon_1}^{(1)} \cap V_{\varepsilon_1}^{(2)} = \emptyset$.

---

*) Here we take

$$S(v,u) = f'(v) - ((f'(v),u)/(g'(u),u))g'(u) .$$

Let $u_0 \in V_{\varepsilon_1}^{(1)}$. Because of $\varepsilon_1 \leqq \varepsilon_0$ it holds that $u_0 \in W_{\varepsilon_0}$ and one obtains from (3) for $t = t_1$

$$f(u(t_1)) \geqq c - \varepsilon_1 + \frac{1}{4}\varepsilon_0 t_1 \geqq c + \varepsilon_1 .$$

Therefore the set of those $t \in [0,t_1]$ for which $f(u(t)) = c + \varepsilon_1$ is nonvoid (and closed) and we denote its greatest lower bound by t^*. On the other hand, from (3) it follows that

$$c + \varepsilon_1 - f(u_0) = f(u(t^*)) - f(u_0) \geqq \frac{1}{4}\varepsilon_0 t^* ,$$

i. e. t^* becomes arbitrarily small if $u_0 \in V_{\varepsilon_1}^{(1)}$ lies sufficiently near (in the sense of the topology on ∂M_R induced by the norm topology) to the level surface $\{u \mid u \in \partial M_R , f(u) = c + \varepsilon_1\}$.

We now displace the element $u_0 \in V_{\varepsilon_1}^{(1)}$ along the trajectory $u(t)$ with the initial point $u(0) = u_0$ and with $h = h_{u_0}$ to the point $u(t^*)$. The points in $V_{\varepsilon_1}^{(2)}$ remain fixed. Therefore, by

$$\varphi(u_0,s) = u(st^*) , \quad \text{if} \quad u_0 \in V_{\varepsilon_1}^{(1)} ,$$

$$= u_0 , \quad \text{if} \quad u_0 \in V_{\varepsilon_1}^{(2)}$$

$(0 \leqq s \leqq 1)$ a continuous deformation of V_{ε_1} in ∂M_R is defined.

If we set $V_{\varepsilon_1}^* = \varphi(V_{\varepsilon_1},1)$ then it holds that $V_{\varepsilon_1}^* \in [V]$ and hence

$$c = \sup_{[V]} \inf_{V} f(u) \geqq \inf_{V_{\varepsilon_1}^*} f(u) \geqq c + \varepsilon_1 .$$

This contradiction proves the lemma, q. e. d.

We again consider the class $[V]$ defined above, and formulate to our variational problem (*) the "complementary" variational problem

$$(**) \qquad \bar{c} = \bar{c}(R) = \inf_{[V]} \sup_{V} f(u) .$$

Similarly as above it holds in the case of f bounded from above on at least one $V \in [V]$ that

$$\inf_{[V]} \sup_{V} f(u) = \inf \{b \in R^1 \mid \exists V \in [V] \text{ with } V \subseteq N_b\} ,$$

where $N_b = f^{-1}(-\infty,b]$.

It is well-known that for the eigenvalues of a (compact) selfadjoint operator in a Hilbert space there are complementary characterizations*). However, in this connection, it must be pointed out that the "complementarity" of (*) and (**) is not directly comparable with the linear case.

For arbitrary $\varepsilon > 0$ we define analogously

$$\overline{W}_\varepsilon = \{u \mid u \in \partial M_R \, , \; |f(u) - \overline{c}| \leqq \varepsilon\}$$

and the following lemma associated to Lemma 3 holds:

Lemma 3´. Let all suppositions of Lemma 3 be fulfilled. Then for each $\varepsilon > 0$ there exists (at least) one $\overline{u}_\varepsilon \in \overline{W}_\varepsilon$ with $\|Q\overline{u}_\varepsilon\| < \varepsilon$.

Proof of Lemma 3´. In the same way as in the proof of Lemma 3 we suppose, contrary to our assertion, that there exists an $\overline{\varepsilon}_0 > 0$ with $\|Qu\| \geqq \overline{\varepsilon}_0$ for all $u \in \overline{W}_{\overline{\varepsilon}_0}$.

Let $u_0 \in \overline{W}_{\overline{\varepsilon}_0}$ be chosen arbitrarily. Then we can find $\overline{h} = \overline{h}_{u_0} \in X$ with $\|\overline{h}\| = 1$ such that $-(Qu_0,\overline{h}) \geqq \frac{1}{2}\|Qu_0\|$ ($\overline{h} = -h$ can be taken with h from the proof of Lemma 3). Then $(Qu_0,\overline{h}) \leqq -\frac{1}{2}\overline{\varepsilon}_0$. Further, there exists $\overline{\delta}_0 > 0$ such that $\|f'(u) - f'(u_0)\| \leqq \overline{\varepsilon}_0/(8c_3)$ holds for all u with $\|u - u_0\| \leqq \overline{\delta}_0$ uniformly for all $u_0 \in \partial M_R$.

We set

$$\overline{t}_1 = \min\left\{t_0, \frac{\overline{\delta}_0}{c_2}, \frac{\overline{\varepsilon}_0}{8c_2c_3}\right\}$$

(c_2 according to Lemma 2, c_3 according to Lemma 1 (ii)). Denoting by $u(t)$ the solution of

$$\frac{du(t)}{dt} = \overline{h}_{u_0} - \frac{(g'(u(t)),\overline{h}_{u_0})}{(g'(u(t)),u(t))}\,u(t) \;, \quad u(0) = u_0$$

for $t \in [0,\overline{t}_1]$ and carrying out a consideration completely analogous to that in the proof of Lemma 3 we obtain

$$\|S(\hat{u}_t,u(s)) - Qu_0\| \leqq \tfrac{1}{4}\overline{\varepsilon}_0 \;, \quad s \in [0,\overline{t}_1] \;.$$

---

*) See e. g. W. Stenger: On the variational principles for eigenvalues for a class of unbounded operators. J. Math. Mech. 17 (1968), 641 - 648.

The equality (2) now yields

$$(3') \qquad f(u(t)) - f(u_0) \leqq -\frac{1}{4}\,\bar{\varepsilon}_0 t$$

for all $t \in [0,\bar{t}_1]$ uniformly for all $u_0 \in \bar{W}_{\bar{\varepsilon}_0}$.

Let $\bar{\varepsilon}_1 = \min(\bar{\varepsilon}_0, \frac{1}{8}\bar{\varepsilon}_0\bar{t}_1)$. As above, for this $\bar{\varepsilon}_1$ we find $V_{\bar{\varepsilon}_1} \in [V]$ with $\sup_{V_{\bar{\varepsilon}_1}} f(u) \leqq c + \bar{\varepsilon}_1$ and, in the present case, we set

$$V^{(1)}_{\bar{\varepsilon}_1} = \{u \mid u \in V_{\bar{\varepsilon}_1} ,\ \bar{c} - \bar{\varepsilon}_1 < f(u) \leqq \bar{c} + \bar{\varepsilon}_1\} ,$$

$$V^{(2)}_{\bar{\varepsilon}_1} = \{u \mid u \in V_{\bar{\varepsilon}_1} ,\ f(u) \leqq \bar{c} - \bar{\varepsilon}_1\} .$$

Let $u_0 \in V^{(1)}_{\bar{\varepsilon}_1}$ be arbitrary. It is $u_0 \in \bar{W}_{\bar{\varepsilon}_0}$, and from (3′) for $t = \bar{t}_1$ we can conclude

$$f(u(\bar{t}_1)) \leqq \bar{c} + \bar{\varepsilon}_1 - \frac{1}{4}\bar{\varepsilon}_0\bar{t}_1 \leqq \bar{c} - \bar{\varepsilon}_1 .$$

This implies that the set of all $t \in [0,\bar{t}_1]$ with $f(u(t)) = \bar{c} - \bar{\varepsilon}_1$ is nonvoid (and closed) and we denote by t^{**} its greatest lower bound (indeed, if we start at $t = 0$ the functional f decreases while it moves along the trajectory $u(t)$ from $u(0) = u_0$ to $u(\bar{t}_1)$, see (3′)). Further we obtain $t^{**} \to +0$ if $u_0 \to f^{-1}(\bar{c} - \bar{\varepsilon}_1)$. Finally, the points in $V^{(2)}_{\bar{\varepsilon}_1}$ remain fixed.

Thus in the same way as in the proof of Lemma 3 it is possible to deform the set $V_{\bar{\varepsilon}_1}$ continuously in ∂M_R . The result of this deformation is denoted by $V^{**}_{\bar{\varepsilon}_1}$. Then we obtain $V^{**}_{\bar{\varepsilon}_1} \in [V]$ and

$$\bar{c} = \inf_{[V]} \sup_{V} f(u) \leqq \sup_{V^{**}_{\bar{\varepsilon}_1}} f(u) \leqq \bar{c} - \bar{\varepsilon}_1 .$$

By this contradiction Lemma 3′ is proved.

4. SOLUTION OF THE EIGENVALUE PROBLEM

After the preliminaries in the last two sections we now focus our attention on the existence of a solution of the eigenvalue problem $\lambda g'(u) = f'(u)$.

Our considerations below refer to the class of operators which map X into X^* and satisfy either the condition of pseudo-monotonicity or the condition $(S)_0$.

(a) A mapping $P : X \to X^*$ is said to be pseudo-monotone if for any sequence $\{u_j\}$ in X with $u_j \rightharpoonup u$ in X and

$$\limsup\,(Pu_j, u_j - u) \leq 0 ,$$

it follows that for each $v \in X$,

$$(Pu, u - v) \leq \liminf\,(Pu_j, u_j - v) .$$

(b) A mapping $P : X \to X^*$ satisfies condition $(S)_0$ if for any sequence $\{u_j\}$ in X with $u_j \rightharpoonup u$ in X , $Pu_j \rightharpoonup w$ in X^* and $(Pu_j, u_j) \to (w,u)$ it follows that $u_j \to u$ in X .

Further we consider operators B which satisfy

Assumption II. The operator B maps X into X^* and satisfies the following conditions:

For any sequence $\{u_j\}$ in X with u_j converging weakly to u in X it follows that

(i) $Bu_j \rightharpoonup Bu$ (weakly in X^*),

(ii) $\lim\,(Bu_j, u_j - u) = 0$.

As in Section 3 we use the deformation invariant class $[V]$ of subsets $V \subset \partial M_R$. The following theorem is the main result of this section.

Theorem 1. Let X be a reflexive Banach space, g and f functionals defined on X . Suppose that ∂M_R is bounded and that g has on X the gradient g' (in the Gateaux sense) which is pseudo-monotone and satisfies assumption I. Further, suppose that f has on X the gradient f' (in the Gateaux sense) which satisfies assumption II. Finally, let f' map bounded sets into bounded sets and for each $\eta > 0$ let there exist $\delta = \delta(\eta) > 0$ such that $\|f'(u) - f'(u_0)\| \leq \eta$ holds for all u with $\|u - u_0\| \leq \delta$ uniformly for all $u_0 \in \partial M_R$.

Then:

(i) There exists a sequence $\{u_j\} \subset \partial M_R$ with

$$\lim f(u_j) = c = \sup_{[V]} \inf_{V} f(u) .$$

(ii) If $\|f'(u_j)\| \geq \text{const} > 0$ for all $j \geq j_0$ (j_0 a certain index, $\{u_j\}$ the sequence from (i)), then there exist $u_0 \in X$ and $\lambda_0 \in R^1$ such that

$$\lambda_0 g'(u_0) = f'(u_0) .$$

<u>Proof of Theorem 1 (i).</u> We choose a sequence of positive numbers ε_j with $\varepsilon_j \to 0$ for $j \to \infty$. Then as a consequence of Lemma 3 we obtain a sequence $\{u_j\} \subset W_{\varepsilon_j}$ (we put briefly $u_{\varepsilon_j} \equiv u_j$) with $\|Qu_j\| < \varepsilon_j$. Clearly $\lim f(u_j) = c$, q. e. d.

<u>Proof of Theorem 1 (ii).</u> Because of the reflexivity of X we can assume without loss of generality that $u_j \rightharpoonup u_0$ in X.

From our assumptions it follows

$$\begin{aligned} 0 < \text{const} &\leq \|f'(u_j)\| \leq \\ &\leq \|Qu_j\| + \frac{\|g'(u_j)\|}{|(g'(u_j),u_j)|} |(f'(u_j),u_j)| \leq \\ &\leq \|Qu_j\| + \text{const } |(f'(u_j),u_j)| , \end{aligned}$$

for $j \geq j_0$, therefore

$$|(f'(u_j),u_j)| \geq \text{const} > 0 \tag{4}$$

for all $j \geq j_1$ (j_1 a suitable index with $j_1 \geq j_0$).

We now define real numbers λ_j by

$$\lambda_j = \frac{(f'(u_j),u_j)}{(g'(u_j),u_j)} , \quad j = 1, 2, \ldots .$$

Obviously it holds that

$$|\lambda_j| \leq \alpha^{-1} |(f'(u_j),u_j)| \leq \text{const}$$

so that we can assume without loss of generality $\lambda_j \to \lambda_0$. Hence $\lambda_0 \neq 0$ since

$$|(f'(u_j),u_j)| = |\lambda_j| \, |(g'(u_j),u_j)| \leq \text{const } |\lambda_j|$$

would imply a contradiction to (4).

Since f' maps bounded sets into bounded sets we obtain for an arbitrary $v \in X$ and sufficiently large j

$$(g'(u_j),u_j - v) =$$
$$= \lambda_j^{-1}(\lambda_j g'(u_j) - f'(u_j),u_j - v) + \lambda_j^{-1}(f'(u_j),u_j - v) =$$
$$= \lambda_j^{-1}(\lambda_j g'(u_j) - f'(u_j),u_j - v) + (\lambda_j^{-1} - \lambda_0^{-1})(f'(u_j),u_j - v) +$$
$$+ \lambda_0^{-1}(f'(u_j),u_j - u_0) + \lambda_0^{-1}(f'(u_j),u_0 - v) \leqq$$
$$\leqq \text{const } \|Qu_j\| + \text{const } |\lambda_j^{-1} - \lambda_0^{-1}| +$$
$$+ \lambda_0^{-1}(f'(u_j),u_j - u_0) + \lambda_0^{-1}(f'(u_j),u_0 - v)$$

and thus by assumption II

$$\limsup\,(g'(u_j),u_j - v) \leqq \lambda_0^{-1}(f'(u_0),u_0 - v) \;. \tag{5}$$

If $v = u_0$ the last inequality provides

$$\limsup\,(g'(u_j),u_j - u_0) \leqq 0 \;.$$

Finally, because of the pseudo-monotonicity of g' (5) yields

$$(g'(u_0),u_0 - v) \leqq \liminf\,(g'(u_j),u_j - v) \leqq$$
$$\leqq \limsup\,(g'(u_j),u_j - v) \leqq \lambda_0^{-1}(f'(u_0),u_0 - v)$$

and since v is arbitrary,

$$g'(u_0) = \lambda_0^{-1}f'(u_0) \;,$$

q. e. d.

If we use Lemma 3$'$ and repeat the same argument as in the proof of Theorem 1, we obtain

Theorem 1$'$. Suppose that all assumptions of Theorem 1 are fulfilled.

Then:

(i) There exists a sequence $\{\bar{u}_j\} \subset \partial M_R$ with

$$\lim f(\bar{u}_j) = \bar{c} = \inf_{[V]} \sup_{V} f(u) \;.$$

(ii) If $\|f'(\bar{u}_j)\| \geqq \text{const} > 0$ for all $j \geqq j_0$ (j_0 a certain index, $\{\bar{u}_j\}$ the sequence from (i)), then there exist $\bar{u}_0 \in X$ and

$\bar{\lambda}_0 \in R^1$ such that

$$\bar{\lambda}_0 g'(\bar{u}_0) = f'(\bar{u}_0) .$$

Remarks. 1. If we suitably strengthen the conditions on f', the inequality $\|f'(u_j)\| \geq \text{const} > 0$ in Theorem 1 (or in Theorem 1′) can be derived. For example, if we have $\Theta \notin \partial M_R$ and assume that f' maps weakly convergent sequences into strongly convergent sequences and that $f'(u) = \Theta$ implies $u = \Theta$ then it holds that $\|f'(u)\| \geq$ $\geq \text{const} > 0$ for all $u \in \partial M_R$ (see [7] and [10], [11], too).

2. In connection with various (complementary) variational characterizations of eigenvalues of a (compact) selfadjoint operator in a Hilbert space it is of interest to study the relation between λ_0, u_0 and $\bar{\lambda}_0$, $\bar{u}_0$ given by the variational characterization in Theorem 1 and Theorem 1′.

3. Strengthening suitably the conditions on g, g' and f' we can pass to the limits $R \to 0$ or $R \to +\infty$. Thus it is possible to derive various bifurcation results for nontrivial solutions of $\lambda g'(u) = f'(u)$ (see [8], [11]).

Theorem 2. Let all suppositions of Theorem 1 be fulfilled, g' satisfying the condition $(S)_0$ instead of the condition of pseudo-monotonicity.

Then:

(i) There exists a sequence $\{u_j\} \subset \partial M_R$ with

$$\lim f(u_j) = c = \sup_{[V]} \inf_{V} f(u) .$$

(ii) If $\|f'(u_j)\| \geq \text{const} > 0$ for all $j \geq j_0$ (j_0 a certain index, $\{u_j\}$ the sequence from (i)), then there exist $u_0 \in X$ and $\lambda_0 \in R^1$ such that

$$\lambda_0 g'(u_0) = f'(u_0) , \quad u_0 \in \partial M_R \;{}^{*)}.$$

We prove only the statement (ii).

Proof of Theorem 2 (ii). As in the proof of Theorem 1 (ii) we obtain a sequence $\{u_j\} \subset \partial M_R$ with $u_j \to u_0$ and $\|Qu_j\| \to 0$ and a sequence $\{\lambda_j\} \subset R^1$ with $\lambda_j \to \lambda_0$, $\lambda_0 \neq 0$.

Let $v \in X$ be arbitrary. For a sufficiently large j it holds

---

*) If f is continuous then $f(u_0) = c$.

that

$$(g'(u_j) - \lambda_0^{-1}f'(u_0),v) = \lambda_j^{-1}(\lambda_j g'(u_j) - f'(u_j),v) + \\ + (\lambda_j^{-1} - \lambda_0^{-1})(f'(u_j),v) + \lambda_0^{-1}(f'(u_j) - f'(u_0),v)$$

and by assumption II (i),

$$g'(u_j) \to \lambda_0^{-1}f'(u_0) \quad \text{in} \quad X^* . \tag{6}$$

On the other hand, one obtains for a sufficiently large j

$$(g'(u_j),u_j) = \lambda_j^{-1}(\lambda_j g'(u_j) - f'(u_j),u_j) + \\ + (\lambda_j^{-1} - \lambda_0^{-1})(f'(u_j),u_j) + \lambda_0^{-1}(f'(u_j),u_j)$$

and the use of assumption II (ii) yields for $j \to \infty$;

$$(g'(u_j),u_j) \to \lambda_0^{-1}(f'(u_0),u_0) ,$$

hence by condition $(S)_0$, $u_j \to u_0$. Since $u_j \in B(u_0,r)$ for all sufficiently large j , (i) of assumption I gives $g'(u_j) \to g'(u_0)$ and together with (6),

$$g'(u_j) \to g'(u_0) = \lambda_0^{-1}f'(u_0) .$$

Finally, since g' is continuous on $B(u_0,r)$, g' is the Fréchet derivative of g on each ball $B(u_0,r')$ with $0 < r' < r$ (see [10]). Therefore, g is continuous and hence $g(u_j) \to g(u_0) = R$, q. e. d.

In the case $g'(\Theta_X) = f'(\Theta_X) = \Theta_{X^*}$, an element $u_* \in X$ is said to be an eigenvector of $\lambda g'(u) = f'(u)$ to the eigenvalue λ_* if

$$\lambda_* g'(u_*) = f'(u_*) , \quad u_* \neq \Theta .$$

If $\Theta \notin \partial M_R$, Theorem 2 yields an existence theorem for an eigenvector and an eigenvalue of $\lambda g'(u) = f'(u)$.

5. EXISTENCE OF AN INFINITE SEQUENCE OF EIGENVECTORS

Before we state the main result of this section we formulate the following

Lemma 4. Let g be a continuous functional on X. Suppose that ∂M_R is bounded and that $\Theta \notin \partial M_R$. Moreover, let

(i) $g(tv) \to +\infty$ for $t \to +\infty$ and $v \in X$ with $\|v\| = 1$, *)

(ii) $g(u) \neq g(t_0 u)$ for arbitrary $u \in \partial M_R$ and arbitrary $t_0 > 1$.

Then ∂M_R and $S = \{u \mid u \in X , \|u\| = 1\}$ are homeomorphic (with respect to the topology which is induced by the norm topology of X).

Proof of Lemma 4. We consider the mapping $T : u \mapsto T(u) = u/\|u\|$ for $u \in \partial M_R$. Since ∂M_R is closed and $\Theta \notin \partial M_R$ we have $\|u\| \geq$ const > 0 for all $u \in \partial M_R$ so that T is well-defined on ∂M_R. Obviously

$$\|T(u_1) - T(u_2)\| \leq \text{const } \|u_1 - u_2\|$$

for arbitrary $u_1, u_2 \in \partial M_R$.

For each $v \in S$, (i) implies the existence of a $\tau > 0$ with $g(\tau v) = R$. If one sets $u = \tau v$ it is $u \in \partial M_R$ and because of $T(u) = v$ the mapping T is onto.

From $u_1, u_2 \in \partial M_R$ with $u_1 \neq u_2$ and $T(u_1) = T(u_2)$ we can conclude without loss of generality that $u_1 = t_0 u_2$ for a certain $t_0 > 1$ which implies $g(u_2) = g(t_0 u_2)$, contrary to (ii). Therefore, for each $v \in S$ there exists exactly one $\tau > 0$ such that $\tau v = u \in \partial M_R$. We define

$$T^{-1}(v) = \tau v \quad \text{with} \quad \tau > 0 \quad \text{and} \quad g(\tau v) = R$$

($v \in S$).

To complete the proof we show the continuity of T^{-1}. Let $\{v_n\} \subset S$ be an arbitrary sequence with $v_n \to v$. Since $v_n \in S$ and $v \in S$, there exist for each $n = 1, 2, \ldots$ exactly one $\tau_n > 0$

---

*) We remark that this condition is not required uniformly with respect to $S = \{u \mid u \in X , \|u\| = 1\}$. Therefore our condition is somewhat weaker than the coerciveness condition $g(v) \to +\infty$ for $\|v\| \to +\infty$.

with $T^{-1}(v_n) = \tau_n v_n$ and exactly one $\tau > 0$ with $T^{-1}(v) = \tau v$. We can extract a convergent subsequence $\{\tau_{n_j}\}$ with $\tau_{n_j} \to \tau^*$. It follows that $\tau^* v \in \partial M_R$ and therefore $\tau^* = \tau$ which implies the convergence of the whole sequence $\{\tau_n\}$ to τ, i. e.

$$T^{-1}(v_n) = \tau_n v_n \to \tau v = T^{-1}(v) ,$$

q. e. d.

Let Y be a topological space.

Definition ([12]). The Ljusternik-Schnirelman category of a subset $N \subset Y$, $\mathrm{cat}(N;Y)$, is the least integer n such that N can be covered by n closed subsets of Y each of which is contractible in Y.

Let g be an even functional on X. We denote by $\widetilde{\partial M_R}$ the space obtained by identifying the antipodal points u and $-u$ of ∂M_R and topologized as in [10]. For the case $g(u) = \frac{1}{2}\|u\|^2$ we obtain $\widetilde{\partial M_R} = P^\infty$, the infinite dimensional projective space. From [10] one can conclude that P^∞ contains sets of arbitrary category. Lemma 4 gives us the possibility to transfer this situation to the space $\widetilde{\partial M_R}$ (we refer the reader to [10], Section 5, where a detailed discussion is given). This fact justifies for each $n = 1, 2, \ldots$ the following definition:

$$[\tilde{V}]_n = \text{class of all subsets } \tilde{V} \subset \widetilde{\partial M_R} \text{ with } \mathrm{cat}(\tilde{V};\widetilde{\partial M_R}) \geq n .$$

We now can reformulate the variational problem (*) (and likewise (**)) on the base of class $[\tilde{V}]_n$ (the functionals f and g are supposed to be even), state the appropriate Lemma 3 as well as Theorem 1 and Theorem 2 (or Lemma 3´ and Theorem 1´, respectively) and carry out the proofs. Then one obtains the following results.

Theorem 3. Let g be a continuous even functional on the reflexive Banach space X with the gradient g' on X. Suppose that g' is pseudo-monotone and satisfies assumption I. Moreover, Let ∂M_R be bounded, $\Theta \notin \partial M_R$ and

$$g(tv) \to +\infty \quad \text{for} \quad t \to +\infty \quad \text{and} \quad v \in S ,$$

$$g(u) \neq g(t_0 u) \quad \text{for arbitrary} \quad u \in \partial M_R \quad \text{and arbitrary} \quad t_0 > 1 .$$

Further, let f be an even functional on X with the gradient f' on X. Suppose that f' satisfies assumption II. Finally, let f' map bounded sets into bounded sets and for each $\eta > 0$ let there exist $\delta = \delta(\eta) > 0$ such that $\|f'(u) - f'(u_0)\| \leq \eta$ holds for all u with $\|u - u_0\| \leq \delta$ uniformly for all $u_0 \in \partial M_R$.

Then it holds for $n = 1, 2, \ldots$:

(i) There exists a sequence $\{u_j^{(n)}\} \subset \partial M_R$ with

$$\lim f(u_j^{(n)}) = \lim f(-u_j^{(n)}) = c_n = \sup_{[\tilde{V}]_n} \inf_{\tilde{V}} \tilde{f}(\tilde{u}) \,.^{*)}$$

(ii) If $\|f'(u_j^{(n)})\| \geq \text{const} > 0$ for the sequence from (i) for all $j \geq j_n$ (j_n a suitable index), then there exist $u_n \in X$ and $\lambda_n \in R^1$ such that

$$\lambda_n g'(u_n) = f'(u_n) \,, \quad \lambda_n g'(-u_n) = f'(-u_n) \,.$$

Theorem 4. Suppose that all assumptions of Theorem 3 are fulfilled, g' satisfying condition $(S)_0$ instead of the condition of pseudo-monotonicity.

Then it holds for $n = 1, 2, \ldots$:

(i) There exists a sequence $\{u_j^{(n)}\} \subset \partial M_R$ with

$$\lim f(u_j^{(n)}) = \lim f(-u_j^{(n)}) = c_n = \sup_{[\tilde{V}]_n} \inf_{\tilde{V}} \tilde{f}(\tilde{u}) \,.$$

(ii) If $\|f'(u_j^{(n)})\| \geq \text{const} > 0$ for the sequence from (i) for all $j \geq j_n$ (j_n a suitable index), then there exist $u_n \in X$ and $\lambda_n \in R^1$ such that

$$\lambda_n g'(u_n) = f'(u_n) \,, \quad u_n \in \partial M_R \,,$$

$$\lambda_n g'(-u_n) = f'(-u_n) \,, \quad (-u_n) \in \partial M_R \,.$$

6. APPLICATION TO NONLINEAR ELLIPTIC EIGENVALUE PROBLEMS

In this section we give some brief remarks on the application of the above results to nonlinear elliptic eigenvalue problems.

On a closed subspace X of the Sobolev space $W_p^m(\Omega)$, where

---

*) If $\tilde{u} \in \tilde{V}$ and $(u,-u) \in V \cup (-V)$ is the corresponding pair of antipodal points then we define $\tilde{f}(\tilde{u}) = f(u) = f(-u)$.

Ω is a bounded domain of the Euclidean space R^n with a sufficiently smooth boundary, we consider functionals g and f. Because of the nature of conditions imposed on g, g' and f in the abstract theory developed above, g and f are assumed in the usual form:

$$g(u) = \int_\Omega G(x,u,\dots,D^m u)\,dx\ ,$$

$$f(u) = \int_\Omega F(x,u,\dots,D^{m-1} u)\,dx$$

with C^1-functions $G(x,\xi)$ and $F(x,\zeta)$. Then the level surface ∂M_R is given by

$$\partial M_R = \left\{ u \mid u \in X\ ,\ \int_\Omega G(x,u,\dots,D^m u)\,dx = R \right\}$$

and we have the Euler-Lagrange expressions corresponding to the functionals g and f:

$$g'(u) = \sum_{|\alpha|\leq m} (-1)^{|\alpha|} D^\alpha G_\alpha(x,u,\dots,D^m u)\ ,$$

$$f'(u) = \sum_{|\beta|\leq m-1} (-1)^{|\beta|} D^\beta F_\beta(x,u,\dots,D^{m-1} u)\ ,$$

where $\alpha = (\alpha_1,\dots,\alpha_n)$ and $\beta = (\beta_1,\dots,\beta_n)$ are multiindices of nonnegative integers ($|\alpha| = \sum_{i=1}^n \alpha_i$, $|\beta| = \sum_{i=1}^n \beta_i$). G_α and F_β are derivatives of G and F : $G_\alpha = \partial G/\partial \xi_\alpha$ and $F_\beta = \partial F/\partial \zeta_\beta$, where $\xi = \{\xi_\alpha \mid |\alpha| \leq m\}$ and $\zeta = \{\zeta_\beta \mid |\beta| \leq m-1\}$.

Under suitable polynomial growth conditions on the functions G_α and F_β, g and f are well-defined on X. In the papers [3], [4], [5] and [10] conditions on G_α and F_β are formulated which imply the necessary properties of g, g' and f' for the application of the abstract results (see [3], [5] for detailed proofs).

In this manner we obtain the existence of a countable sequence of eigenvectors and eigenvalues of the problem

$$\lambda \sum_{|\alpha|\leq m} (-1)^{|\alpha|} D^\alpha G_\alpha(x,u,\dots,D^m u) =$$

$$= \sum_{|\beta|\leq m-1} (-1)^{|\beta|} D^\beta F_\beta(x,u,\dots,D^{m-1} u)\ .$$

With respect to our discussion in the appendix we make the following obvious remark: Under appropriate coerciveness conditions on

$G(x,\xi)$ the level surface ∂M_R is bounded in the norm of $W_p^m(\Omega)$, i. e. ∂M_R is bounded in X $(X \subseteq W_p^m(\Omega))$. However, if, conversely, we deal with the variational problems (*) or (**) for the functional g (instead of f) on $\{u \mid u \in X , f(u) = R\}$, the boundedness of the level surface in the norm of $W_p^m(\Omega)$ can no longer be expected.

APPENDIX. VERIFICATION OF CONDITION (C)

Let M be a C^1-Banach manifold. We denote by $T_u(M)$ the tangential space at the point $u \in M$.

The manifold M is called a Finsler manifold if $T(M)$ (the tangential bundle of M) has a Finsler structure, i. e. roughly speaking, for each point $u_0 \in M$ there is a neighborhood (the domain of a suitable bundle chart) such that the norms of all $T_u(M)$ with u in this neighborhood are equivalent (the norms vary continuously on this neighborhood). By means of the Finsler structure on $T(M)$, a metric on M can be defined. This metric is consistent with the topology of M ([12]).

Let $h : M \to R^1$ be a C^1-function on the manifold M . Then $h'(u)$ is a bounded linear functional on $T_u(M)$. If $h'(u) \neq \theta$ then the point u is called a regular point of h . The number $c \in R^1$ is called a regular value of f if $h^{-1}(c)$ contains only regular points.

Let M be a C^1-Finsler manifold, h a C^1-function on M .

Condition (C) (Palais-Smale). If $\{u_j\} \subset M$ is a sequence of points such that $h(u_j)$ is bounded and $h'(u_j)$ converges to zero, then $\{u_j\}$ contains a convergent subsequence.

Let X be a Banach space. We can regard X in a trivial way as a C^k-manifold $(k \geq 1)$. Further, let f be a C^1-function on X and R a regular value of f . Then according to the smoothness theorem for regular levels,*) the level surface $M(f) = f^{-1}(R) = \{u \mid u \in X , f(u) = R\}$ is a closed C^1-submanifold of X . Therefore $M(f)$ is a complete Finsler manifold in the Finsler structure induced by the

*) R. S. Palais: Morse theory on Hilbert manifolds. Topology 2 (1963), 299 - 340.

flat Finsler structure on the product bundle $X \times X = T(X)$ ([12]). The tangential space at the point $u \in M(f)$ is defined by

$$T_u(M(f)) = \{v \mid v \in X\ ,\ (f'(u),v) = 0\}\ .$$

<u>Lemma.</u> Let f be a C^1-function on the reflexive Banach space X whose derivative f' maps bounded sets into bounded sets and satisfies condition II of Section 4. Suppose that $|(f'(u),u)| \geq c = $ $= \text{const} > 0$ for all $u \in M(f)$.

Further, let g be a C^1-function which is coercive on X: $|g(u)| \to \infty$ for $\|u\| \to \infty$. Suppose that g' maps bounded sets into bounded sets and satisfies condition $(S)_0$.

Then g satisfies condition (C) on $M(f)$.

<u>Proof of Lemma.</u> Let $\{u_j\} \subset M(f)$ be an arbitrary sequence with $|g(u_j)| \leq \text{const}$ and $\|g'(u_j)\|_{T^*_{u_j}(M)} \to 0$.

Since g is coercive we have $\|u_j\| \leq \text{const}$ and therefore one can assume without loss of generality that $u_j \rightharpoonup u_0$ in X.

Let $z \in X$ with $\|z\| = 1$ be arbitrarily chosen. Then for all $u \in M(f)$ the vector $v = z - ((f'(u),z)/(f'(u),u))u$ is an element of $T_u(M(f)) = T_u(M)$. Denoting $Qu = g'(u) -$ $- ((g'(u),u)/(f'(u),u))f'(u)$ for $u \in M(f)$, we obtain

$$(g'(u),v) = (Qu,z)\ . \tag{7}$$

We now consider the sequence $\{u_j\}$. Since f' maps bounded sets into bounded sets, the estimation

$$\|v_j\| \leq 1 + \frac{\|f'(u_j)\|}{|(f'(u_j),u_j)|} \leq c_0 = \text{const}$$

where

$$v_j = z - \frac{(f'(u_j),z)}{(f'(u_j),u_j)}\, u_j$$

holds for all $j = 1, 2, \ldots$. If $v_j = \Theta$ we obtain from (7) that $(Qu_j,z) = 0$. On the other hand, in the case $v_j \neq \Theta$ we get by means of the above estimation

$$|(Qu_j,z)| \leq c_0 \left|\left(g'(u_j), \frac{v_j}{\|v_j\|}\right)\right| \leq$$

$$\leq c_0 \sup_{v \in T_{u_j}(M),\ \|v\|_{T_{u_j}(M)}=1} |(g'(u_j),v)| = c_0 \|g'(u_j)\|_{T^*_{u_j}(M)}\ .$$

Since $z \in X$ with $\|z\| = 1$ was arbitrary it follows that

$$\|Qu_j\|_{X^*} = \sup_{z \in X, \|z\|_X = 1} |(Qu_j, z)| \leq c_0 \|g'(u_j)\|_{T^*_{u_j}(M)} \to 0 .$$

To complete the proof we use the technique of the proof of Theorem 2 (ii). We set $\mu_j = (g'(u_j), u_j)/(f'(u_j), u_j)$. Then $|\mu_j| \leq$ const and we can assume without loss of generality that $\mu_j \to \mu_0$.

For arbitrary $z \in X$ it holds that

$$|(g'(u_j) - \mu_0 f'(u_0), z)| \leq \|Qu_j\| \|z\| +$$

$$+ |\mu_j - \mu_0| |(f'(u_j), z)| + |\mu_0| |(f'(u_j) - f'(u_0), z)| .$$

Thus $g'(u_j) \to \mu_0 f'(u_0)$ in X^* with regard to assumption II. Further

$$|(g'(u_j), u_j) - (\mu_0 f'(u_0), u_0)| \leq$$

$$\leq \|Qu_j\| \|u_j\| + |\mu_j - \mu_0| |(f'(u_j), u_j)| +$$

$$+ |\mu_0| |(f'(u_j), u_j - u_0)| + |\mu_0| |(f'(u_j) - f'(u_0), u_0)| ,$$

i. e. $(g'(u_j), u_j) \to (\mu_0 f'(u_0), u_0)$. Therefore, condition $(S)_0$ now implies $u_j \to u_0$, q. e. d.

The conditions of our above lemma are somewhat more closely related to nonlinear elliptic eigenvalue problems than those in [5], Part II. A more detailed discussion of the approach to nonlinear eigenvalue problems which is sketched in the present section will be given by the author in a forthcoming paper.

REFERENCES

[1] M. S. Berger: An eigenvalue problem for non-linear elliptic partial differential equations. Trans. Amer. Math. Soc. 120 (1965), 145 - 184.

[2] M. S. Berger: A Sturm-Liouville theorem for nonlinear elliptic partial differential equations. Ann. Scuola Norm. Sup. Pisa Serie 3 20 (1966), 543 - 582.

[3] F. E. Browder: Infinite dimensional manifolds and non-linear elliptic eigenvalue problems. Ann. Math. 72 (1965), 459 - 477.

[4] F. E. Browder: Nonlinear eigenvalue problems and Galerkin approximations. Bull. Amer. Math. Soc. 74 (1968), 651 - 656.

[5] F. E. Browder: Existence theorems for nonlinear partial differential equations. To appear.

[6] E. S. Citlanadze: Existence theorems for min-max-points in Banach spaces and their applications (Russian). Trudy mosk. matem. obšč. 2 (1953), 235 - 274.

[7] S. Fučík, J. Nečas: Ljusternik-Schnirelmann theorem and nonlinear eigenvalue problems. To appear in Math. Nachr.(1972).

[8] M. A. Krasnoselskij: Topological methods in the theory of nonlinear integral equations (Russian). Moscow 1956.

[9] L. A. Ljusternik: Topology of functional spaces and calculus of variations in the large (Russian). Trudy mat. inst. V. A. Steklova 19 (1947).

[10] J. Naumann: Lusternik-Schnirelman-Theorie und nichtlineare Eigenwertprobleme. To appear.

[11] J. Naumann: Variationsmethoden für Existenz und Bifurkation von Lösungen nichtlinearer Eigenwertprobleme II. To appear.

[12] R. S. Palais: Lusternik-Schnirelman theory on Banach manifolds. Topology 5 (1966), 115 - 132.

[13] M. M. Vajnberg: Variational methods for the investigation of nonlinear operators (Russian). Moscow 1956.

Humboldt-Universität zu Berlin, Sektion Mathematik, Unter den Linden 6, 108 Berlin, GDR

FREDHOLM THEORY OF BOUNDARY VALUE PROBLEMS FOR NONLINEAR ORDINARY DIFFERENTIAL OPERATORS

JINDŘICH NEČAS, PRAHA (CZECHOSLOVAKIA)

INTRODUCTION

(i) Nonlinear boundary value problems for both ordinary and partial differential operators lead to the study of a nonlinear equation $Tu - \lambda Su = f$, where T and S are operators from a real reflexive Banach space B to the dual space B^* , and $\lambda \in R_1$. The operator T plays the part of the identity mapping and S is completely continuous. Below, we shall define asymptotes of operators T and S , namely T_0 and S_0 . The asymptotes are $\varkappa$-homogeneous: $T_0(tu) = t^{\varkappa}Tu$, $t > 0$, $\varkappa > 0$ and analogously for S_0 .

The following situation occurs: $T - \lambda S$ is onto if (and in a sense iff) λ is not an eigenvalue of the equation $T_0u - \lambda S_0u = 0$.

This is the "Fredholm theory". There is a nice open problem: if $T - \lambda S$ is not onto, what conditions guarantee $(T - \lambda S)(B)$ to be closed? This problem is equivalent in practical cases to the so called normal solvability of $Tu - \lambda Su = f$.

"Fredholm theory" or Fredholm alternative was discovered and studied in papers by S. I. Pochožajev [1], J. Nečas [2], [3], F. E. Browder [4], M. Kučera [5], S. Fučík [6], R. I. Kačurovskij [7].

(ii) If we suppose T and S to be gradient operators of potentials f and g then, using the ideas of L. A. Ljusternik, L. Schnirelmann, we can prove the existence of a countable number of eigenvalues with ∞ as the limit. This subject was considered in many papers; from the more recent ones we mention e. g. F. E. Browder [4], E. S. Citlanadze [8], S. Fučík, J. Nečas [9], J. Naumann [10]; in these papers there is a large list of references of anterior works. Especially, we must not forget the books by M. M. Vajnberg [11], M. A. Krasnoselskij [12], M. S. Berger, M. S. Berger [13] and lecture notes by J. T. Schwartz [14]. The theory of Ljusternik-Schnirelmann having been developed in connection with the study of differential and integral equations, it is applicable to those equations.

If we neglect some obvious situations, of the same type as for instance to solve the "eigenvalue problem" for the equation $x - \lambda x^2 = 0$, a question arises how to characterize the eigenvalues different from those obtained by Ljusternik-Schnirelmann.

(iii) Therefore, in the nonlinear Fredholm theory, we have to deal with at least a countable number of eigenvalues. Is the real situation worse and how much? Using Sard's theorem in the case of a finite dimensional space, S. I. Pochožajev tried to answer this question in his paper [15]. On the other hand, if we consider real analytic functionals f and g, it is proved in papers of S. Fučík, J. Nečas, J. Souček, V. Souček [24], [25], [26] that the number of critical values of g with respect to the manifold $f(u) = c$ is at most countable. This requires, for example in Hilbert space, the second differential of f to be positive definite.

The study of norms as potentials leads to the eigenvalue problem for duality mappings. So we arrive at a nonlinear Sturm-Liouville problem for the equation

$$-(a(x)|y'(x)|^{p-2}\, y'(x))' + (b(x) - \lambda c(x))|y(x)|^{p-2}\, y(x) = 0 .$$

It is shown in the paper by J. Nečas [16] for this equation and in the paper by A. Kratochvíl, J. Nečas [17] for the fourth-order equation that all normed eigenvectors are isolated and therefore countable and that all eigenvalues are $0 < \lambda_1 < \lambda_2 < \dots$ with $\lambda_n \to \infty$.

The situation for higher order operators remains obscure; evidently, we cannot expect eigenvectors to be isolated in general.

Anyway, the Rayleigh's quotient for the nonlinear Sturm-Liouville problem is only twice differentiable and critical vectors are precisely the eigenvectors. This together with the example of I. Kupka [18] (see also A. Marino, S. Spagnolo [19]) points to the exclusiveness of second-order (fourth-order) differential operators.

This is a brief outline of the present lecture notes.

CHAPTER 1. FREDHOLM THEORY

§ 1. Ordinary differential operators.

If $1 \leqq p < \infty$ and k is a natural number, we denote by $W_p^{(k)}$ the Sobolev space of absolutely continuous real functions u on the interval $\langle 0,1 \rangle$ whose derivatives up to the order $k-1$ are also absolutely continuous and whose derivative of the order k is L_p function. $W_p^{(k)}$ is a Banach space with the norm

$$\|u\|_{k,p} \overset{df}{=} \left(\sum_{i=0}^{k} \int_0^1 |u^{(i)}(x)|^p \, dx \right)^{\frac{1}{p}} . \tag{1.1}$$

Banach spaces $C^{(k)}$ of k times continuously differentiable functions and $C^{(k),\mu}$ of Hölder-continuous functions in the interval $\langle 0,1 \rangle$ are introduced as usual.

In the sequel we shall consider only the case $1 < p < \infty$.

If $y \in W_p^{(k)}$, let us define $\xi(y) \in [L_p]^{1+k}$ by $\xi(y) = (y, y^{(1)}, \dots, y^{(k)})$ and $\eta(y) \in [L_p]^k$ by $\eta(y) = (y, y^{(1)}, \dots, y^{(k-1)})$.

Let for $i = 0,1,\dots,k$, $a_i(x,\xi)$ be a continuous function of $\langle 0,1 \rangle \times R_{1+k}$ into R_1 (the Caratheodory condition would suffice, too). Put $\overset{0}{W}{}_p^{(k)} = \{u \in W_p^{(k)} \mid u(0) = u(1) = \dots = u^{(k-1)}(0) = u^{(k-1)}(1) = 0\}$.

Let V be a subspace of $W_p^{(k)}$ determined by ℓ linearly independent conditions

$$\sum_{j=0}^{k-1} (C_{ij}^0 y^{(j)}(0) + C_{ij}^1 y^{(j)}(1)) = 0 , \tag{1.2}$$

$1 \leqq i \leqq \ell \leqq 2k$, where C_{ij}^s $(i = 1,\dots,\ell$, $j = 0,\dots,k-1$, $s = 0, 1)$ are real constants ($\ell = 0$ means that no conditions are given).

Let $A_i^0(\eta)$, $A_i^1(\eta)$, $i = 0,1,\dots,k-1$ be continuous functions from R_k to R_1. The weak differential operator $T : W_p^{(k)} \to V^*$ is defined by

$$(Tu,v) \overset{df}{=} \int_0^1 \sum_{i=0}^{k} a_i(x,\xi(u)(x))v^{(i)}(x) \, dx + \tag{1.3}$$

$$+ \sum_{i=0}^{k-1} (A_i^0(\eta(u)(0))v^{(i)}(0) + A_i^1(\eta(u)(1))v^{(i)}(1)) ,$$

where $(\ ,\ )$ denotes the natural pairing between V^* and V.

We shall assume

$$(1.4) \qquad |a_k(x,\xi)| \leq C(|\eta|)\,(1 + |\xi_k|^{p-1})$$

and, for $i < k$:

$$(1.5) \qquad |a_i(x,\xi)| \leq C(|\eta|)\,(1 + |\xi_k|^p) ,$$

where $C(s)$ is a positive continuous function.

Further, let $b_i(x,\eta)$, $i = 0,1,2,\ldots,k$ be continuous functions in $\langle 0,1\rangle \times R_k$ and $B_i^0(\eta)$, $B_i^1(\eta)$, $i = 0,1,\ldots,k-1$ continuous functions in R_k .

The weak differential operator $S : W_p^{(k)} \to V^*$ is defined by

$$(1.6) \qquad (Su,v) \overset{df}{=} \int_0^1 \sum_{i=0}^{k} b_i(x,\eta(u)(x))v^{(i)}(x)\,dx + \\ + \sum_{i=0}^{k-1} (B_i^0(\eta(u)(0))v^{(i)}(0) + B_i^1(\eta(u)(1))v^{(i)}(1)) .$$

In all cases we shall suppose

$$(1.7) \qquad (\xi_k' - \xi_k)\,(a_k(x,\eta,\xi_k') - a_k(x,\eta,\xi_k)) > 0$$

for $\xi_k' \neq \xi_k$ and

$$(1.8) \qquad \xi_k a_k(x,\xi) \geq c_1|\xi_k|^p - c(|\eta|)$$

with $c_1 > 0$.

We can also consider some other conditions, for example

$$(1.9) \qquad \sum_{i=0}^{k} \xi_i\, a_i(x,\xi) \geq c|\xi|^p ,$$

$$(1.10) \qquad \sum_{i=0}^{k-1} (A_i^0(\eta)\eta_i + A_i^1(\eta')\eta_i') \geq 0$$

for η , η' satisfying (1.2).

As for the asymptote of T , we can further suppose that there exist continuous functions $\tilde{a}_i(x,\xi)$, $i = 0,1,\ldots,k$, (p-1)-homogeneous in ξ : $\tilde{a}_i(x,t\xi) = t^{p-1}\tilde{a}_i(x,\xi)$, $t > 0$, such that

$$(1.11) \qquad \left|\frac{a_i(x,t\xi)}{t^{p-1}} - \tilde{a}_i(x,\xi)\right| \leq \omega(t)(c(|\eta|) + |\xi_k|^{p-1}) ,$$

with $\lim_{t\to\infty} \omega(t) = 0$, and there exist (p-1)-homogeneous continuous functions $\tilde{A}_i^0(\eta)$, $\tilde{A}_i^1(\eta)$ such that

$$(1.12) \qquad \left|\tilde{A}_i^0(\eta) - \frac{A_i^0(t\eta)}{t^{p-1}}\right| \le \omega(t)\ c(|\eta|)$$

and analogously for $\tilde{A}_i^1$.

The functions $\tilde{a}_i(x,\xi)$ as well as $\tilde{A}_i^0(\eta)$, $\tilde{A}_i^1(\eta)$ are supposed to be odd in ξ.

In all cases when $\tilde{a}_k(x,\xi)$ is considered, we suppose

$$(1.13) \qquad (\xi_k' - \xi_k)\ (\tilde{a}_k(x,\eta,\xi_k') - \tilde{a}_k(x,\eta,\xi_k)) > 0$$

for $\xi_k' \neq \xi_k$ and

$$(1.14) \qquad \xi_k\ \tilde{a}_k(x,\xi) \ge c_1\,|\xi_k|^p - c(|\eta|)\ .$$

Concerning the asymptote of S, we can assume the existence of functions $\tilde{b}_i(x,\eta)$, $\tilde{B}_i^0(\eta)$, $\tilde{B}_i^1(\eta)$ which are odd, $(p-1)$-homogeneous and satisfy

$$(1.15) \qquad \left|\frac{b_i(x,t\eta)}{t^{p-1}} - \tilde{b}_i(x,\eta)\right| \le \omega(t)\ c(|\eta|)\ ,$$

$$(1.16) \qquad \left|\tilde{B}_i^j(\eta) - \frac{B_i^j(t\eta)}{t^{p-1}}\right| \le \omega(t)\ c(|\eta|)\ ,\ j = 0,\ 1\ .$$

Let numbers s_i, $i = 1,2,\ldots,\ell$ and n_j^0, n_j^1, $j = 0,1,\ldots,k-1$ and a function $f \in L_1$ be given. (This can be generalized in obvious directions; for example if $V = \overset{o}{W}{}_p^{(k)}$, then instead of $f \in L_1$ we can take $f_i \in L_p^*$, $i = 0,1,\ldots,k$ and replace the functional $\int_0^1 f(x)v(x)\,dx$ by $\sum_{i=0}^{k} \int_0^1 f_i(x)v^{(i)}(x)\,dx$.)

A weak solution u of the boundary value problem, satisfying, on the boundary, the stable conditions

$$(1.17) \qquad \sum_{j=0}^{k-1} (C_{ij}^0 u^{(j)}(0) + C_{ij}^1 u^{(j)}(1)) = s_i \quad (i = 1,2,\ldots,\ell)$$

is a function u which fulfills

$$(1.18) \qquad \int_0^1 \sum_{i=0}^{k} (a_i(x,\xi(u)(x)) - \lambda b_i(x,\eta(u)(x)))v^{(i)}(x)\,dx +$$

$$+ \sum_{i=0}^{k-1} \Big[(A_i^0(\eta(u)(0)) - \lambda B_i^0(\eta(u)(0)))v^{(i)}(0) +$$

$$+ (A_i^1(\eta(u)(1)) - \lambda B_i^1(\eta(u)(1)))v^{(i)}(1)\Big] =$$

$$= \int_0^1 f(x)v(x)\,dx + \sum_{i=0}^{k-1} (n_i^0\, v^{(i)}(0) + n_i^1\, v^{(i)}(1))$$

for each $v \in V$.

§ 2. Fredholm alternative

In the sequel, B is a reflexive, separable Banach space. We consider $T : B \to B^*$ and $S : B \to B^*$. Let the following conditions be fulfilled:

(2.1) T is demicontinuous, i. e. if $u_n \to u$ (strong convergence), then $Tu_n \rightharpoonup Tu$ (weak convergence),

(2.2) T is bounded, i. e. T maps bounded sets in B onto bounded sets in B^* ,

(2.3) S is completely continuous,

(2.4) T satisfies condition (S): if $u_n \rightharpoonup u$ and $(Tu_n - Tu, u_n - u) \to 0$, then $u_n \to u$.

It is possible to impose some other conditions on T , S :

(2.5) (coerciveness) $$\lim_{\|u\|\to\infty} \frac{((T - \lambda S)u,u)}{\|u\|} = \infty .$$

Theorem 2.1. If conditions (2.1) - (2.5) are satisfied, then $(T - \lambda S)(B) = B^*$ and $(T - \lambda S)^{-1}$ (in general a multivalued mapping) is bounded.

For the proof see F. E. Browder [4] or with obvious modifications the proof of the next theorem (where the notion of the degree is not necessary).

Let us try to substitute condition (2.5) by a weaker one:

(2.6) (weak coerciveness) $$\lim_{\|u\|\to\infty} \|(T - \lambda S)u\| = \infty ;$$

on the other hand let us suppose that a demicontinuous, bounded operator $R : B \to B^*$ exists such that

(2.7) $$T - \lambda S - R \quad \text{is odd,}$$

(2.8) $$\lim_{\|u\|\to\infty} \frac{\|Ru\|}{\|(T - \lambda S)u\|} = 0 ,$$

and

(2.9) for $T - \tau R$, $0 \leq \tau \leq 1$, condition (S) is satisfied.

Theorem 2.2. If conditions (2.1) - (2.4), (2.6) - (2.9) are satisfied, then $(T - \lambda S)(B) = B^*$ and $(T - \lambda S)^{-1}$ is bounded (as a multivalued mapping).

Proof. Let $B = \bigcup_{n=1}^{\infty} B_n$, $\dim B_n = n$, $B_1 \subset B_2 \subset \dots$. If F is a finite dimensional subspace of B , let ψ_F be the identity mapping $F \to B$ where the norm in F is an equivalent one. Put $(u^*, \psi_F u) \overset{df}{=} (\psi_F^* u^*, u)$ for $u^* \in B^*$, $u \in F$. Hence $\psi_F^* : B^* \to F^*$. Put $\psi_n \overset{df}{=} \psi_{B_n}$ and let $f \in B^*$. Define for each natural n and $t \in \langle 0,1 \rangle$ $A_{n,t}$: $B_n \to B_n^*$ as follows:

$$A_{n,t} u \overset{df}{=} \psi_n^* (T - \lambda S - (1 - t)R) \psi_n u - t \psi_n^* f .$$

Let us identify B_n and B_n^* with R_n provided with the usual scalar product.

Proposition. There exists $\varrho > 0$ and n_0 such that if $\|u\| = \varrho$, $u \in B_n$ and $n \geq n_0$, then $A_{n,t} u \neq 0$ for $0 \leq t \leq 1$.

First, it is obvious that

$$\lim_{\|u\| \to \infty} \|(T - \lambda S - (1 - t)R)u - tf\| = \infty$$

uniformly with respect to $t \in \langle 0,1 \rangle$. Hence there exists $\varrho > 0$ such that if $\|u\| = \varrho$, $0 \leq t \leq 1$ then $\|(T - \lambda S - (1 - t)R)u - tf\| > 0$.

Let us suppose that there exists $u_{n_k} \in B_{n_k}$, $\|u_{n_k}\| = \varrho$, $n_k \to \infty$ and t_k such that $A_{n_k, t_k} u_{n_k} = 0$. We can suppose $t_k \to t$, $u_{n_k} \rightharpoonup u$ (because of the reflexivity of the space B).

Let $v_{n_k} \in B_{n_k}$ such that $v_{n_k} \to u$. Then

$$0 = (A_{n_k, t_k} u_{n_k}, u_{n_k} - v_{n_k}) = (A_{t_k} u_{n_k}, u_{n_k} - v_{n_k})$$

and therefore

$$\lim_{k \to \infty} (A_t u_{n_k} - A_t u, u_{n_k} - u) = 0 .$$

Because of the complete continuity of S , A_t obviously satisfies condition (S), hence $u_{n_k} \to u$; therefore $\|u\| = \varrho$ and $A_t u = 0$, which is impossible. Since $A_{n,t} u \neq 0$ for $\|u\| = \varrho$, $u \in B_n$, $A_{n,t}$ is continuous and the identity mapping maps the set $\{u \in B_n \mid \|u\| < \varrho\}$ onto a symmetric, open subset D_n of R_n , we have $\text{degree}(A_{n,t}, 0, D_n) = \text{degree}(A_{n,0}, 0, D_n) \neq 0$ by the homotopy argument of Brouwer degree and the Borsuk-Ulam theorem on odd mappings. Therefore there exist $u_n \in B_n$ such that $\|u_n\| < \varrho$ and $A_{n,1}u_n = 0$ which is equivalent to

(2.10) $\forall v \in B_n : ((T - \lambda S)u_n - f, v) = 0$ (Galerkin's method).

We can suppose $u_n \rightharpoonup u$, $v_n \in B_n$, $v_n \to u$ which yields

$$\lim_{n\to\infty} ((T - \lambda S)u_n - (T - \lambda S)u, u_n - u) =$$

$$= \lim_{n\to\infty} ((T - \lambda S)u_n, u_n - v_n) = \lim_{n\to\infty} (f, u_n - v_n) = 0 ,$$

hence $u_n \to u$ and from (2.10) it follows finally that $(T - \lambda S)u = f$ and $\|u\| \leq \varrho$, q. e. d.

In order to find sufficient conditions for the weak coerciveness, let us suppose that T has a $\varkappa$-asymptote at the infinity: there exists a bounded, demicontinuous, $\varkappa$-homogeneous and odd operator T_0 such that

$$\lim_{\|u\|\to\infty} \frac{\|(T_0 - T)(u)\|}{\|u\|^{\varkappa}} = 0 . \tag{2.11}$$

Let us suppose that S has also a $\varkappa$-asymptote at the infinity: there exists a completely continuous $\varkappa$-homogeneous and odd operator S_0 such that

$$\lim_{\|u\|\to\infty} \frac{\|(S - S_0)u\|}{\|u\|^{\varkappa}} = 0 . \tag{2.12}$$

A real number λ is called an eigenvalue for the operator $T_0 - \lambda S_0$ iff there exists $u \neq 0$ such that $(T_0 - \lambda S_0)u = 0$. The element u is said to be an eigenvector. We say that $(T - \lambda S)^{-1}$ is $\varkappa$-bounded, if there exists $C > 0$ such that

$$\|u\| \leq C(1 + \|(T - \lambda S)u\|)^{\frac{1}{\varkappa}} \quad \text{for each} \quad u \in B . \tag{2.13}$$

Theorem 2.3. Let conditions (2.1) - (2.3), (2.11) - (2.12) be valid and let $tT + (1 - t)T_0$ for $0 \leq t \leq 1$ satisfy condition (S).

Then λ is not an eigenvalue for $T_0 - \lambda S_0$ iff $(T - \lambda S)(B) = B^*$ and $(T - \lambda S)^{-1}$ is $\varkappa$-bounded.

Proof. Suppose λ is not an eigenvalue. Then

$$\|u\| \leq c\,\|(T_0 - \lambda S)u\|^{\frac{1}{\varkappa}} ; \tag{2.14}$$

indeed, in virtue of the $\varkappa$-homogeneity of T_0, S_0 it is sufficient to show that $\inf_{\|u\|=1} \|(T_0 - \lambda S_0)u\| > 0$. If not, then $\exists u_n$, $\|u_n\| = 1$, $u_n \rightharpoonup u$, $\|(T_0 - \lambda S_0)u_n\| \to 0$, $Su_n \to u^*$. Condition (S) implies then $u_n \to u$. Thus $(T_0 - \lambda S_0)u = 0$ which is impossible.

Relations (2.11), (2.12) and (2.14) imply (2.6) and putting $R = T - \lambda S - T_0 + \lambda S_0$, we have (2.8) and (2.9) and we can apply Theorem 2.2.

Suppose λ is an eigenvalue. u being an eigenvector for $T_0 - \lambda S_0$, we have

$$\lim_{t\to\infty} \frac{\|(T - \lambda S)tu\|}{t^{\varkappa}} = 0 ,$$

therefore (2.13) is not true, q. e. d.

§ 3. Boundary value problems

The only nonobvious fact which is to be verified is condition (S); we do it for the coefficient a_i, the proof for the coefficient $ta_i + (1 - t)\tilde{a}_i$ following precisely the same lines. The reader should notice that all members cancelled from T form a completely continuous operator and therefore do not affect condition (S).

Lemma 3.1. Let $a_k(x,\xi)$ satisfy conditions (1.4), (1.7), (1.8). If $u_n \rightharpoonup u$ in $W_p^{(k)}$, $\omega \in W_p^{(k)}$ and

$$(3.1) \qquad \int_0^1 (a_k(x,\xi(\omega)(x) + \xi(u_n)(x)) - a_k(x,\xi(\omega)(x) + \xi(u)(x)))(u_n^{(k)}(x) - u^{(k)}(x))\,dx \to 0 ,$$

then $u_n \to u$ in $W_p^{(k)}$.

Proof. In virtue of the fact that the identity mapping from $W_p^{(1)}$ to C is compact, it follows from (3.1):

$$(3.2) \qquad \int_0^1 (a_k(x,\xi(\omega)(x) + \xi(u_n)(x)) - a_k(x,\eta(\omega)(x) + \eta(u_n)(x),\omega^{(k)}(x) + u^{(k)}(x)))(u_n^{(k)}(x) - u^{(k)}(x))\,dx \to 0 .$$

Put

$$f_n(x) \overset{df}{=} (a_k(x, \xi(\omega)(x) + \xi(u_n)(x) - a_k(x, \eta(\omega)(x) +$$

$$+ \eta(u_n)(x), \omega^{(k)}(x) + u^{(k)}(x)))(u_n^{(k)}(x) - u^{(k)}(x)) .$$

Because of (1.7), $f_n(x) \geq 0$ and we can suppose $f_n(x) \to 0$ almost everywhere. Condition (1.4) together with (1.8) show that $u_n^{(k)}(x)$ is a bounded sequence almost everywhere and in virtue of (1.7), every subsequence is necessarily convergent to $u^{(k)}(x)$ therefore the same is true for $u_n^{(k)}(x)$. On the other hand, conditions (1.4), (1.8) together with (3.2) show the equicontinuity of

$$\int_M |u_n^{(k)}(x)|^p \, dx .$$

Now it follows from Vitali's theorem $u_n^{(k)} \to u^{(k)}$ in L_p , q. e. d.

Theorem 3.1. Let the operators T and S be given as in § 1 and let (1.4), (1.5), (1.7), (1.8), (1.11) - (1.16) be satisfied. Then, if the only solution $u \in V$ of the homogeneous problem

$$\int_0^1 \sum_{i=0}^{k} (\tilde{a}_i(x, \xi(u)(x)) - \lambda \tilde{b}_i(x, \xi(u)(x)))v^{(i)}(x) \, dx +$$

$$+ \sum_{i=0}^{k-1} \Big[(\tilde{A}_i^0(\eta(u)(0)) - \lambda \tilde{B}_i^0(\eta(u)(0)))v^{(i)}(0) +$$

$$+ (\tilde{A}_i^1(\eta(u)(1)) - \lambda \tilde{B}_i^1(\eta(u)(1)))v^{(i)}(1) \Big] = 0$$

(for each $v \in V$) is zero, there exists a solution of the problem (1.17), (1.18) with $s_i = 0$ and for every such solution u we have

$$\|u\|_{k,p} \leq c(1 + \sum_{i=0}^{k-1} (|n_i^0| + |n_i^1|) + \|f\|_{0,1})^{\frac{1}{p-1}} . \tag{3.3}$$

If, in addition, $\tilde{a}_i$, $\tilde{b}_i$, $\tilde{A}_i^0$, $\tilde{A}_i^1$, $\tilde{B}_i^0$, $\tilde{B}_i^1$ fulfill conditions (written only for $\tilde{a}_i$)

$$2 \leq p < \infty : \quad |\tilde{a}_i(x,\xi) - \tilde{a}_i(x,\xi')| \leq$$

$$\leq C(1 + |\xi| + |\xi'|)^{p-2} |\xi - \xi'| ,$$

$$1 < p < 2 : \quad |\tilde{a}_i(x,\xi) - \tilde{a}_i(x,\xi')| \leq C|\xi - \xi'|^{p-1} ,$$

then there exists a solution of the problem (1.17), (1.18) and for every such solution u we have

$$\|u\|_{k,p} \leq C(1 + \sum_{i=1}^{\ell} |s_i| + \sum_{i=0}^{k-1} (|n_i^0| + |n_i^1|) + \|f\|_{0,1})^{\frac{1}{p-1}} .$$

CHAPTER 2. LJUSTERNIK-SCHNIRELMANN THEORY

§1. Potentials of differential operators

In this chapter, we apply abstract theorems to gradients of potentials.

We consider a continuous function $F(x,\xi)$ in $\langle 0,1\rangle \times R_{k+1}$, even in ξ, $F(x,0) = 0$, satisfying

$$|F(x,\xi)| \leq c(|\eta|)(1 + |\xi_k|)^p , \tag{1.1}$$

which is continuously differentiable in ξ_j and such that

$$\left|\frac{\partial F}{\partial \eta_i}\right| \leq c(|\eta|)(1 + |\xi_k|)^p , \tag{1.2}$$

$$\left|\frac{\partial F}{\partial \xi_k}\right| \leq c(|\eta|)(1 + |\xi_k|)^{p-1} .$$

Further, we suppose

$$(\xi_k' - \xi_k)\left(\frac{\partial F}{\partial \xi_k}(x,\eta,\xi_k') - \frac{\partial F}{\partial \xi_k}(x,\eta,\xi_k)\right) > 0 \tag{1.3}$$

for $\xi_k' \neq \xi_k$ and

$$\sum_{i=0}^{k} \frac{\partial F}{\partial \xi_i}(x,\xi)\xi_i \geq c_1|\xi_k|^p + c_2|\eta|^p , \quad c_1 > 0 , \quad c_2 \geq 0 \tag{1.4}$$

and $c_2 > 0$ if the space $V \neq \overset{o}{W}{}_p^{(k)}$ is considered. We also define continuously differentiable even functions $H_0(\eta)$, $H_1(\eta)$ from R_k into R_1, $H_0(0) = H_1(0) = 0$, such that

$$(1.5) \qquad \sum_{i=0}^{k-1} \frac{\partial H_0}{\partial \eta_i}\eta_i \geq 0 , \quad \sum_{i=0}^{k-1} \frac{\partial H_1}{\partial \eta_i}\eta_i \geq 0 .$$

A functional $f : W_p^{(k)} \to R_1$ will be considered:

$$(1.6) \qquad f(u) = \int_0^1 F(x,\xi(u)(x))\, dx + H_0(\eta(u)(0)) + H_1(\eta(u)(1)) .$$

Further we consider a continuous even in η function $G(x,\eta)$ from $\langle 0,1\rangle \times R_k$ into R_1 , $G(x,0) = 0$, G being continuously differentiable in η_j and such that

$$(1.7) \qquad \sum_{i=0}^{k-1} \frac{\partial G}{\partial \eta_i}(x,\eta)\eta_i > 0 \quad \text{for} \quad \eta \neq 0 .$$

Further, let functions $N_0(\eta)$, $N_1(\eta)$ with the same properties as H_0 , H_1 be given.

We define $g : W_p^{(k)} \to R_1$ by

$$(1.8) \qquad g(u) = \int_0^1 G(x,\eta(u)(x))\, dx + N_0(\eta(u)(0)) + N_1(\eta(u)(1)) .$$

If $1 < p < 2$, then the functions F , H_0 , H_1 , G , N_0 , N_1 are supposed to satisfy the following conditions written only for F :

$$(1.9) \qquad \left|\frac{\partial F}{\partial \eta_i}(x,\xi) - \frac{\partial F}{\partial \eta_i}(x,\xi')\right| \leq c(|\eta| + |\eta'|)|\xi - \xi'|^{p-1}$$

and if $p \geq 2$

$$(1.10) \qquad \left|\frac{\partial F}{\partial \eta_i}(x,\xi) - \frac{\partial F}{\partial \eta_i}(x,\xi')\right| \leq$$

$$\leq c(|\eta| + |\eta'|)(1 + |\xi_k|^{p-2} + |\xi_k'|^{p-2})|\xi - \xi'| .$$

We shall look for the eigenvalues of the operator $f' - \lambda g'$ (where f' and g' denote Frechet derivatives of f and g , respectively), considering the critical values of $g(u)$ on the manifold $\{u | f(u) = c\}$.

§ 2. Ljusternik's theorem

In the abstract language, this chapter deals with some even functionals f and g from the reflexive separable Banach spaces B to the reals. We shall suppose

(2.1) f and g have a Frechet derivative,

(2.2) f' and g' are locally uniformly continuous,

(2.3) g' is strongly continuous, i. e. if $u_n \rightharpoonup u$, then $g'(u_n) \to g'(u)$,

(2.4) f' satisfies condition (S),

(2.5) $f(u) = 0 \Leftrightarrow u = 0$,

(2.6) $g(u) = 0 \Leftrightarrow u = 0$.

Let us denote $M_c(f) = \{u \in B \mid f(u) = c\}$. The tangent space to $M_c(f)$ at a point u is

$$T_u(M) = \{h \in B \mid (f'(u),h) = 0\} .$$

The manifold $M_c(f)$ is called regular if

$$f'(u) \neq 0 \quad \text{for each} \quad u \in M_c(f) .$$

A point $u \in M_c(f)$ is a critical point of a functional g with respect to $M_c(f)$, if for each $h \in T_u(M)$ we have $(g'(u),h) = 0$. The value $g(u) = \gamma$ is called the critical value.

Lemma 2.1. Let f and g have Frechet derivatives and let $M_c(f)$ be a regular manifold. If u is a critical point of g with respect to $M_c(f)$, then there exists $\mu \in R_1$ such that $g'(u) = \mu f'(u)$.

Proof. There exists $v_0 \in B$ such that $(f'(u),v_0) = 1$. Every point $w \in B$ can be expressed uniquely in the form $w = c(w)v_0 + v$ with $v \in T_u(M)$ because $c(w) = (f'(u),w)$. However,

$$(g'(u),w) = c(w)(g'(u),v_0) + (g'(u),v) = \mu(f'(u),w) ,$$

where $\mu = (g'(u),v_0)$.

§ 3. Order of sets

Instead of introducing and working with the notion of the category of a compact set, we shall introduce the notion of the order of a set which is in fact very close to that just mentioned and seems to be more suitable to deal with. The definition is a modification of the definition due to M. A. Krasnoselskij [12].

Let K be a closed set in a Banach space B. Iff $K = \emptyset$, we put $\text{ord } K = 0$. If $K \neq \emptyset$ and $u \in K \Rightarrow -u \notin K$, then we put $\text{ord } K = 1$. $\text{ord } K = n$, if there exist closed subsets of K, say K_i, $i = 1,2,\ldots,n$, such that $\text{ord } K_i = 1$, $K = \bigcup_{i=1}^{n} K_i$ and n is minimal. We put $\text{ord } K = \infty$ if it is not of finite order.

Let us denote by S_n the unit sphere in R_n.

Lemma 3.1. $\text{ord } S_n = n + 1$.

Proof. Obviously, $\text{ord } S_n \leq n + 1$. Let us suppose $\text{ord } S_n \leq n$. Then there exist $K_i \subset S_n$, $i = 1,2,\ldots,n$ such that $K_i \cap -K_i = \emptyset$ and $\bigcup_{i=1}^{n} K_i = S_n$. Put

$$\zeta_i(x) = 1 \quad \text{for } x \in K_i ,$$

$$= -1 \quad \text{for } x \in -K_i$$

and let us extend $\zeta_i(x)$ continuously on S_n. Put $\zeta(x) = (\zeta_1(x),\ldots,\zeta_n(x))$. ζ is a continuous mapping of S_n to R_n and the theorem that every continuous mapping $S_n \to R_{n-1}$ has an antipodal point x, i. e. $f(x) = f(-x)$ yields a contradiction: $\bar{\zeta}(x) = \zeta(x)/|\zeta(x)|$ as a mapping from S_n to S_n omits the point $(1,0,\ldots,0)$. Putting $\bar{\bar{\zeta}}(x) = \pi\bar{\zeta}(x)$ where π is the stereographic projection with the point $(1,0,\ldots,0)$ as North Pole, it is $\bar{\bar{\zeta}}(x_0) = \bar{\bar{\zeta}}(-x_0)$ for some point x_0 which is not possible.

Lemma 3.2. Let K be a compact set in B_1 and A a continuous odd mapping of K into B_2. Then $\text{ord } AK \geq \text{ord } K$.

Proof. If $\text{ord } AK = \infty$ the assertion is true. Let $\text{ord } AK = 1$ and let us suppose $\text{ord } K > 1$. Then there exists a point $u \in K$ such that $-u \in K$. However, then also $Au \in AK$ as well as $-Au \in AK$, which is impossible. If $\text{ord } AK = n$, then $AK = \bigcup_{i=1}^{n} D_i$, $\text{ord } D_i = 1$. Put $H_i = A^{-1}D_i$. We have proved $\text{ord } H_i = 1$, hence $\text{ord } K \leq n$, q. e. d.

The following properties of the order are evident:

(3.1) $$K_1 \subset K_2 \Rightarrow \text{ord } K_1 \leq \text{ord } K_2 ,$$

(3.2) $$\text{ord } (K_1 \cup K_2) \leq \text{ord } K_1 + \text{ord } K_2 .$$

Lemma 3.3. For a compact set K, there exists a neighbourhood $U(K)$ of K such that $\text{ord } \overline{U(K)} = \text{ord } K$.

§ 4. Ljusternik-Schnirelmann theorem

We obtain without any difficulty

Lemma 4.1. Let f be an even nonnegative functional satisfying (2.5) and such that for its Frechet derivative

(4.1) $$(f'(u),u) > 0 \quad \text{for} \quad u \neq 0 .$$

Further, let

(4.2) $$\lim_{\|u\| \to \infty} f(u) = \infty .$$

Then every half-line from the origin intersects $M_c(f)$ precisely in one point and this mapping is a homeomorphism of the unit sphere S onto $M_c(f)$.

Let V_k be the set of all compact subsets $K \subset M_c(f)$ such that $\text{ord } K \geq k$.

For a functional $g(u)$ and the family of sets V_k, put

$$\hat{g}(K) = \inf_{u \in K} g(u) , \qquad \gamma_k = \sup_{K \in V_k} \hat{g}(K) .$$

Let us recall: a Banach space is called locally uniformly convex if $\|x_n\| \to \|x\|$ and $\|x_n + x\| \to 2\|x\|$ implies $\|x_n - x\| \to 0$.

Let us define the duality mapping $J : B^* \to 2^B$ by

$$Ju^* \overset{df}{=} \{u \in B \mid (u^*,u) = \|u^*\|^2 , \|u\| \leq \|u^*\|\} .$$

Lemma 4.2. If B is a reflexive and locally uniformly convex space, then J is a singlevalued mapping and it is continuous.

Proof. As B is locally uniformly convex, it follows immediately that every convex set on the unit sphere is precisely one point. But Ju^* is a convex set and $\|u^*\|^2 = (u^*,u) \leq \|u^*\| \|u\|$, hence $\|u\| = \|u^*\|$ and therefore Ju^* is singlevalued. If $u_n^* \to u^*$, then

$\|Ju_n^*\| = \|u_n^*\| \to \|u^*\| = \|Ju^*\|$. But from the definition of Ju^* it follows that $Ju_n^* \rightharpoonup Ju^*$, hence $\underline{\lim} \|Ju_n^* + Ju^*\| \geq 2\|Ju^*\|$ and $\lim_{n\to\infty} \|Ju_n^* + Ju^*\| = 2\|Ju^*\|$ which implies $Ju_n^* \to Ju^*$.

<u>Lemma 4.3</u>. (Fundamental.) Let f and g be even nonnegative functionals satisfying (2.1), (2.2), (2.5), (2.6), (4.1), (4.2) and

(4.3) if $c > 0$, then $\inf_{u \in M_c(f)} (f'(u),u) = c_1 > 0$.

(4.4) f' and g' are bounded operators.

Put

$$Au \stackrel{df}{=} g'(u) - \frac{(g'(u),u)}{(f'(u),u)} f'(u)$$

for $u \in M_c(f)$. Let k be fixed. Then for each α , $0 < \alpha \leq 1$, there exists $\varepsilon \in (0,\alpha)$ such that if $K \in V_k$ and $\hat{g}(K) > \gamma_k - \varepsilon$, then there exists $u \in K$ such that $|g(u) - \gamma_k| < \alpha$ and $\|Au\| < \alpha$.

<u>Proof</u>. S. L. Troyanski proved in the paper [23] that every reflexive space B has an equivalent norm for which both B and B^* are locally uniformly convex. Hence we can suppose B and B^* to be such spaces. Put for $u \in M_c(f)$

$$p(u) \stackrel{df}{=} JAu - \frac{(f'(u),JAu)}{(f'(u),u)} .$$

Then p is an odd, continuous and bounded mapping from $M_c(f)$ to B . For ε , $t \in R_1$, $u \in M_c(f)$, let us define $\varphi(u,t,\varepsilon) \stackrel{df}{=} f(u + tp(u) + \varepsilon u) - c$. We have $\varphi(u,0,0) = 0$, φ is continuous on $M_c(f) \times R_2$ with continuous derivatives $\frac{\partial\varphi}{\partial t}$, $\frac{\partial\varphi}{\partial\varepsilon}$. We have $\frac{\partial\varphi}{\partial\varepsilon}(u,0,0) = (f'(u),u) \geq c_1 > 0$. The implicit functions theorem implies the existence of δ , $0 < \delta < \frac{1}{2}$ such that for $|t| < \delta$ there exists a unique $\varepsilon(u,t)$, $|\varepsilon(u,t)| \leq c_2\delta$ such that $\varphi(u,t,\varepsilon(u,t)) = 0$ and $\varepsilon(u,t)$, $\frac{\partial\varepsilon}{\partial t}(u,t)$ are continuous in $M_c(f) \times (-\delta,\delta)$. We have

$$\frac{\partial\varepsilon}{\partial t}(u,t) = -\frac{\frac{\partial\varphi}{\partial t}}{\frac{\partial\varphi}{\partial\varepsilon}}$$

and $\frac{\partial\varphi}{\partial t}(u,0,0) = 0$.

Let us suppose that the assertion of Lemma 4.3 is not true. Put $\varepsilon = \alpha^2\delta/8$. Therefore there exists $K \subset V_k$, $\hat{g}(K) > \gamma_k - \varepsilon$, such that for $u \in K \cap \{u \mid |g(u) - \gamma_k| < \alpha\}$ it is $\|Au\| \geq \alpha$. Put $L = \{u \in K \mid g(u) < \gamma_k + \alpha\}$ and for $u \in M_c(f)$,

$$\varkappa(u) = u + \delta p(u)/2 + \varepsilon(u,\delta/2)u \ .$$

For $u \in L$ we have

$$g(\varkappa(u)) = g(u) + \\ + \int_0^{\delta/2} (g'(u + tp(u) + \varepsilon(u,t)u), p(u) + \frac{\partial\varepsilon}{\partial t}(u,t)u)\, dt \ .$$

But $(g'(u),p(u)) = (Au,JAu) = \|Au\|^2 \geq \alpha^2$, hence for δ small enough and $u \in K \cap L$, the inequality $g(\varkappa(u)) \geq \gamma_k + \varepsilon$ holds. For $u \in K \setminus L$, if δ is small enough, $\varkappa(u)$ is so close to u that $g(\varkappa(u)) \geq \gamma_k + \alpha/2 = \gamma_k + \varepsilon$. Hence $\hat{g}(\varkappa(K)) \geq \gamma_k + \varepsilon$ and since ord $\varkappa(K) \geq k$, we obtain a contradiction, q. e. d.

Theorem 4.1. Let for f and g the assumptions of Lemma 4.3 be fulfilled, let g' satisfy also condition (2.3) and

$$g'(u) = 0 \Leftrightarrow u = 0 \ . \tag{4.5}$$

Let f' fulfill condition (2.4).

Then for every natural number k there exists $u_k \in M_c(f)$ such that $g(u_k) = \gamma_k$ and

$$g'(u_k) - \frac{(g'(u_k),u_k)}{(f'(u_k),u_k)} f'(u_k) = 0 \ . \tag{4.6}$$

It is $\gamma_k \geq \gamma_{k+1} > 0$.

Proof. As $0 \notin M_c(f)$, it is $g(u) > 0$ for $u \in K \in V_k$ and, K being compact, it follows that $\hat{g}(K) > 0$, hence $\gamma_k > 0$. The relation $V_{k+1} \subset V_k$ implies $\gamma_{k+1} \leq \gamma_k$. According to the fundamental lemma, there exists $u_n \in M_c(f)$ such that $\lim_{n\to\infty} g(u_n) = \gamma_k$ and $\lim_{n\to\infty} \|Au_n\| = 0$. We can suppose $u_n \rightharpoonup u_k$ and by (2.3), $g(u_k) = \gamma_k$, hence $u_k \neq 0$. We can suppose

$$\frac{(g'(u_n),u_n)}{(f'(u_n),u_n)} \to \mu$$

and $\mu \neq 0$ by (4.5). Hence $f'(u_n) \rightharpoonup g(u_k)/\mu$ and condition (2.4) implies $u_n \to u_k$, therefore $u_k \in M_c(f)$ and (4.6) is valid, q. e. d.

The above Theorem 4.1 does not contain any assertion on the existence of a countable number of γ_k . If the Banach space under consideration has some convenient geometrical structure of the type "Schauder basis", such an assertion is really true.

A Banach space B has the usual structure, if there exists a sequence of operators (in general, nonlinear) $P_n : B \to B$, such that

(i) P_n is continuous,

(ii) P_n is odd,

(iii) the linear hull $[P_n(B)]$ is of a finite dimension,

(iv) $u_n \to u \Rightarrow P_n u_n \to u$.

In the paper by S. Fučík, J. Milota [20], the authors proved that if P_n are linear and B reflexive, then (iv) is equivalent to $P_n^* x^* \to x^*$.

<u>Theorem 4.2.</u> Let the space B have the usual structure. Let f and g satisfy the assumptions of Theorem 4.1. Then $\gamma_n \to 0$, $u_n \to 0$ and $\lambda_n \to \infty$ where

$$\lambda_n = \frac{(f'(u_n), u_n)}{(g'(u_n), u_n)} .$$

<u>Proof.</u> Let $\varepsilon > 0$. Then it is easy to see that there exists $n_0 > 0$ such that

$$|g(P_{n_0} u) - g(u)| < \frac{\varepsilon}{2}$$

for $u \in M_c(f)$ and

$$g(u) < \frac{\varepsilon}{2} \quad \text{for} \quad \|u\| \leq \varrho .$$

Hence if $u \in M_c(f)$ and $\|P_{n_0} u\| \leq \varrho$, then $g(u) < \varepsilon$. If $K \subset M_c(f) \cap \cap \{u \mid g(u) \geq \varepsilon\}$ then $\|P_{n_0} u\| > \varrho$ for $u \in K$. But ord $P_{n_0}(K) \leq \leq \dim [P_{n_0}(B)] + 1$. If $k \geq \dim [P_{n_0}(B)] + 2$ and $K \in V_k$, it must exist $u \in K$ such that $\|P_{n_0} u\| \leq \varrho$ and therefore $\hat{g}(K) < \varepsilon$ and $\gamma_k \leq \varepsilon$. Consequently, $\gamma_k \to 0$. If $u_k \not\to 0$, then there is a subsequence u_{k_j}, $u_{k_j} \to u \neq 0$, but $g(u_{k_j}) = \gamma_{k_j} \to 0$ and $g(u) = 0 \Rightarrow \Rightarrow u = 0$, which is a contradiction. Because of the assumption (2.3) it is $\lambda_n \to \infty$, q. e. d.

<u>Remark.</u> Actually, to obtain γ_k, u_k in Theorems 4.1 and 4.2, it is sufficient to define V_k as follows ($k \geq 2$):

If B_{k-1} is a subspace of B of the dimension $k - 1$, then V_k consists of all sets $M_c(f) \cap B_{k-1}$ and all sets $A(M_c(f) \cap B_{k-1})$ where A is an odd mapping $M_c(f) \cap B_{k-1} \to M_c(f)$.

What happens if $\gamma_k = \gamma_{k+1} = \dots = \gamma_{k+p}$?

Theorem 4.3. Let the assumptions of Theorem 4.1 be fulfilled and let

$$\gamma_k = \gamma_{k+1} = \dots = \gamma_{k+p} = \gamma .$$

Let $T_0 = \{u \in M_c(f) \mid Au = 0 , g(u) = \gamma\}$ (Au being from Lemma 4.3). Then $\operatorname{ord} T_0 \geq 1 + p$.

Proof. Clearly, T_0 is compact. Let

$$T_\alpha = \{u \in M_c(f) \mid \|Au\| \leq \alpha , |g(u) - \gamma| \leq \alpha\} .$$

Let us suppose $\operatorname{ord} T_0 \leq p$. According to Lemma 3.3, $\operatorname{ord} \overline{U(T_0,\eta)} \leq p$ for some $U(T_0,\eta)$. If $\alpha_0 > 0$ is small enough, then $T_\alpha \subset \overline{U(T_0,\eta/2)}$ for $\alpha \in \langle 0,\alpha_0\rangle$. Let $\varepsilon(\alpha/2)$ be that from Lemma 4.3. If $K \in V_{k+p}$ and $\gamma - \varepsilon(\alpha/2) < \hat{g}(K) \leq \gamma$, then $k + p \leq \operatorname{ord} (K \setminus U(T_\alpha,\eta/2)) + \operatorname{ord}(K \cap \overline{U(T_\alpha,\eta/2)})$ and since $\operatorname{ord} (K \cap \overline{U(T_\alpha ,\eta/2)}) \leq p$, we obtain $K \setminus U(T_\alpha,\eta/2) \in V_k$. Therefore

$$\gamma \geq \inf_{u \in K \setminus U(T_\alpha,\eta/2)} g(u) \geq \inf_{u \in K} g(u) > \gamma - \varepsilon ,$$

hence there exists $u \in K \setminus U(T_\alpha,\eta/2)$ such that $|g(u) - \gamma| < \alpha/2$ and $\|Au\| < \alpha/2$ and so $u \in T_{\alpha/2}$ which is impossible, q. e. d.

It is easy to see that compact sets T_0 , that means the sets of critical points of g for the same critical value, are not too big:

Theorem 4.4. Let the assumptions of Theorem 4.2 hold and let T_0 be the set from the preceding theorem. Then there exists an integer-valued function $n(k)$ such that $\operatorname{ord} T_0 < n(k)$.

Proof. Let $0 < \varepsilon < \gamma_k$ as in the proof of Theorem 4.2. Put $n(k) = 2 + \dim [P_{n_0}(B)]$. Let us suppose $\operatorname{ord} T_0 \geq n(k)$. Then $\hat{g}(T_0) \leq \varepsilon$, hence $\gamma_k \leq \varepsilon$ which is a contradiction, q. e. d.

Remark. If f and g from Theorem 4.1 are $(\varkappa+1)$-homogeneous, then $\lambda_n = c/\gamma_n$ because

$$g(u_n) = \int_0^1 (g'(tu_n),u_n)\, dt = \frac{(g'(u_n),u_n)}{\varkappa + 1}$$

and

$$f(u_n) = \int_0^1 (f'(tu_n),u_n)\, dt = \frac{(f'(u_n),u_n)}{\varkappa + 1} .$$

§ 5. Application to differential operators

First we must verify the assumptions for f and g : (2.1) is obvious; (2.2) follows from (1.9) or (1.10); (2.3) is obvious; (2.4) follows from (1.3), (1.4) as in Lemma 3.1, Chap. 1; (2.5) follows from (1.4), (1.5); (2.6) follows from (1.7) and (1.5) for H_0, H_1; (4.1) is a consequence of (1.4) and (1.5); (4.2) follows from (1.4) and (1.5); (4.3) follows from (1.4) and (1.5); (4.4) is obvious; (4.5) follows from (1.7) and (1.5) for H_0, H_1.

We shall prove that the spaces V have the usual structure.

We shall obtain a little more; it is possible to show as in the paper by S. Fučík, O. John, J. Nečas [21] that V has a Schauder basis; nevertheless proving the usual structure is essential and easy for ordinary differential operators.

Let $f_0, f_1, \dots, f_k \in L_p$, $1 < p < \infty$ and let us look for a function $u \in V_p$ such*) that for each $v \in V_{p*}$:

$$\text{(5.1)} \qquad \int_0^1 \sum_{i=0}^{k} u^{(i)} v^{(i)} \, dx = \int_0^1 \sum_{i=0}^{k} f_i v^{(i)} \, dx .$$

We have

Lemma 5.1. There exists a unique solution of (5.1) and the operator $G : [f_0, \dots, f_k] \to u$ is linear and bounded.

Proof. Let us suppose $f_i \in C$. The problem (5.1) can be written in the form

$$\text{(5.2)} \qquad (u,v)_k = \sum_{i=0}^{k} (f_i, v^{(i)})_0 \quad \text{for each } v \in V_2 ,$$

where $(\ ,\ )_j$ is a scalar product in $W_2^{(j)}$. According to the Riesz Theorem there exists a unique $u \in V_2$. However, (5.2) implies for $v \in W_2^{(k)}$ by integration by parts:

$$\text{(5.3)} \qquad \int_0^1 \Big(u^{(k)}(x) - f_k(x) +$$

$$+ \sum_{i=0}^{k-1} \Big[(-1)^{k-i} \frac{1}{(k-i-1)!} \int_0^x (x-\xi)^{k-i-1} \, (u^{(i)}(\xi) -$$

---

*) $V_p \equiv V$ where the index p accents that V is the subspace of $W_p^{(k)}$.

$$- f_i(\xi))\, d\xi \Big] v^{(k)}(x)\Big)\, dx = 0 \ .$$

Put

$$M(x) = u^{(k)}(x) - f_k(x) +$$

$$+ \sum_{i=0}^{k-1} (-1)^{k-i} \frac{1}{(k - i - 1)!} \int_0^x (x - \xi)^{k-i-1} (u^{(i)}(\xi) - f_i(\xi))\, d\xi \ .$$

There exist uniquely determined constants $c_0, c_1, \dots, c_{k-1}$ such that

$$\int_0^1 (M(x) - c_0 - c_1 x - \dots - c_{k-1} x^{k-1}) x^j\, dx = 0 \ ,$$

$j = 0,1,\dots,k-1$ and evidently

$$(5.4) \qquad |c_i| \leq C(\|u\|_{k-1,1} + \sum_{i=0}^{k} \|f_i\|_{0,1} +$$

$$+ |u^{(k-1)}(0)| + |u^{(k-1)}(1)|) \ .$$

Hence instead of $v^{(k)}(x)$ in (5.3), we can put there any $g \in L_p$ and we obtain

$$(5.5) \qquad \|u\|_{k,p} \leq c(\|u\|_{k-1,1} + \sum_{i=0}^{k} \|f_i\|_{0,p} +$$

$$+ |u^{(k-1)}(0)| + |u^{(k-1)}(1)|) \ .$$

We have

$$(5.6) \qquad \|u\|_{k-1,1} + |u^{(k-1)}(0)| + |u^{(k-1)}(1)| \leq C \sum_{i=0}^{k} \|f_i\|_{0,p} \ .$$

Indeed, if this were not true, then there would exist a sequence

$$f_i^n \in C \ , \quad \sum_{i=0}^{k} \|f_i^n\|_{0,p} = 1$$

and solutions of (5.2) u_n such that

$$(5.7) \qquad \|u_n\|_{k-1,p} + |u_n^{(k-1)}(0)| + |u_n^{(k-1)}(1)| > n \ .$$

Put $v_n = u_n / \|u_n\|_{k,p}$. Because of (5.7), we then have $\|u_n\|_{k,p} \to \infty$. We can suppose $v_n \rightharpoonup v$, hence it follows from (5.5) that $v_n \to v$ and therefore we have $\|v\|_{k,p} = 1$, $v \in V_p$, and $(v,h)_k = 0$ for each $h \in V_{p^*}$. But it follows from (5.3) that $v \in W_2^{(k)}$, hence $v \equiv$

$= 0$, which is a contradiction. We have also proved that the solution of (5.1) is unique. Therefore we obtain from (5.5), (5.6) for $f_i \in C$:

$$(5.8) \qquad \|G((f_0, \ldots, f_k))\|_{k,p} \leq c \sum_{i=0}^{k} \|f_i\|_{0,p} .$$

However, (5.8) permits to extend the operator G continuously on the whole space $[L_p]^{k+1}$, q. e. d.

Theorem 5.1. In V_p , $1 < p < \infty$, there exist linear P_n : $B \to B$, $\|P_n\| \leq c < \infty$, $\dim P_n(B) \leq n$ and $u_n \to u \Rightarrow P_n u_n \to u$.

Proof. The space $[L_{p^*}]^{k+1}$ having a Schauder basis, there exist $Q_n : [L_{p^*}]^{k+1} \to [L_{p^*}]^{k+1}$ such that $\dim Q_n [L_{p^*}]^{k+1} = n$, $\|Q_n\| \leq c$, $Q_n^2 = Q_n$, and $Q_n u \to u$. Let $M_n = Q_n^*$. We obtain $\|M_n\| \leq c$, $\dim M_n [L_p]^{k+1} = n$ and $u_n \to u \Rightarrow M_n u_n \to u$. Let I be the imbedding $W_p^{(k)}$ in $[L_p]^{k+1}$ defined by $f_i = u^{(i)}$. Put $P_n = GM_n I$. If $u_n \to u$, then $Iu_n \to Iu$, $M_n(Iu_n) \to Iu$, hence $GM_n Iu_n \to GIu = u$, q. e. d.

Theorem 5.2. If conditions (1.1) - (1.10) are satisfied, then for each $c > 0$, there exists an infinite number of $u_n \in V$ such that u_n are critical points for the functional (1.8) with respect to the manifold $f(u) = c$, where $f(u)$ is defined in (1.6). For $g(u_n) = \gamma_n$, we have $\gamma_n \geq \gamma_{n+1}$, $\gamma_n \to 0$, $u_n \to 0$, the eigenvalues λ_n are positive and $\lambda_n \to \infty$.

We can apply also other theorems of the preceding section.

CHAPTER 3. SPECTRUM OF DUALITY MAPPINGS

§ 1. Second order operators

If some positive power of the norm in a Banach space has a continuous Frechet derivative, then for $u \neq 0$:

$$\|u\|^p = p \int_0^1 (\|tu\|^{p-1} \|tu\|',u)\, dt = \|u\|^{p-1} (\|u\|',u) ,$$

hence this derivative represents a duality mapping (or one point in

Ju). Therefore it is natural to consider first the spectrum of such mappings.

Let V be a subspace of $W_p^{(1)}$ given by one of the relations

(1.1) (i) no relation, i. e. $V = W_p^{(1)}$,

(ii) one relation $u(0) = 0$ or $u(1) = 0$,

(iii) $u(0) = u(1) = 0$, i. e. $V = \overset{o}{W}{}_p^{(1)}$,

and let us consider the functional

$$(1.2) \qquad f(u) \overset{df}{=} \frac{1}{p} \left(\int_0^1 [a(x)|u'(x)|^p + b(x)|u(x)|^p] \, dx + \right.$$

$$+ \frac{1}{p} A_0 |u(0)|^p + \frac{1}{p} A_1 |u(1)|^p$$

and the functional

$$(1.3) \qquad g(u) \overset{df}{=} \frac{1}{p} \int_0^1 c(x)|u(x)|^p \, dx +$$

$$+ \frac{1}{p} B_0 |u(0)|^p + \frac{1}{p} B_1 |u(1)|^p .$$

We suppose $a(x) > 0$, $a \in C^{(1)}$, b , $c \in C$, $b(x) \geq 0$, $c(x) > 0$, $A_0 \geq 0$, $A_1 \geq 0$, $B_0 \geq 0$, $B_1 \geq 0$, $p \geq 2$; in the case (i) or (ii) of (1.1) we suppose $b(x) \not\equiv 0$ or $A_0 + A_1 > 0$.

Hence the smallest eigenvalue λ_1 of $f'(u) - \lambda g'(u)$ is positive and we suppose

$$(1.4) \qquad \lambda_1 c(x) - b(x) > 0 .$$

Moreover, in the case (i) we suppose either $B_0 = B_1 = 0$ and $A_0 + A_1 > 0$ or

$$(1.5) \qquad \lambda B_0 - A_0 \geq 0 , \quad \lambda B_1 - A_1 \geq 0 ,$$

$$\lambda B_0 - A_0 + \lambda B_1 - A_1 > 0 .$$

We formulate the main theorem, its proof being based on some lemmas introduced below.

<u>Theorem 1.1</u>. Let conditions (1.1) - (1.5) be satisfied. Then all eigenvalues of $f'(u) - \lambda g'(u)$ form a countable set $0 < \lambda_1 < \lambda_2 < \lambda_3 < \dots$ such that $\lim \lambda_n = \infty$. Eigenvectors are isolated and to each λ_n there exists a finite number of normed eigenvectors.

<u>Remark</u>. If $V = \overset{o}{W}{}_p^{(1)}$, it is easy to prove that to λ_1 there corresponds exactly one pair of normed eigenvectors, u and $-u$,

$u(x) > 0$ in $(0,1)$; v being an eigenvector corresponding to any other eigenvalue, there exists $x_0 \in (0,1)$ such that $v(x_0) = 0$.

The following relations are immediate consequences of the definition:

$$(1.6)\qquad df(u,v) = \int_0^1 (a|u'|^{p-2} u'v' + b|u|^{p-2} uv)\, dx +$$

$$+ A_0|u(0)|^{p-2} u(0)v(0) + A_1|u(1)|^{p-2} u(1)v(1) ,$$

$$dg(u,v) = \int_0^1 c|u|^{p-2} uv\, dx +$$

$$+ B_0|u(0)|^{p-2} u(0)v(0) + B_1|u(1)|^{p-2} u(1)v(1) ,$$

$$(1.7)\qquad d^2f(u,v,h) = (p-1)\int_0^1 (a|u'|^{p-2} v'h' + b|u|^{p-2} vh)\, dx +$$

$$+ (p-1)A_0|u(0)|^{p-2} v(0)h(0) + (p-1)A_1|u(1)|^{p-2} v(1)h(1) ,$$

$$d^2g(u,v,h) = (p-1)\int_0^1 c|u|^{p-2} vh\, dx +$$

$$+ (p-1)B_0|u(0)|^{p-2} v(0)h(0) + (p-1)B_1|u(1)|^{p-2} v(1)h(1) .$$

Before passing to the proof of Theorem 1.1, let us make the following agreement: we give a natural number N and we consider only such eigenvalues that $\lambda_k \leq N$. Concerning the eigenvectors, we consider only those for which $\|u\|_{1,p} \leq 2$.

For $x_0 \in \langle 0,1\rangle$, let us introduce the function

$$(1.8)\qquad M_{x_0}(x) \overset{df}{=} a(x)|u'(x)|^{p-2} u'(x) +$$

$$+ \int_{x_0}^{x} (\lambda c(\xi) - b(\xi))|u(\xi)|^{p-2} u(\xi)\, d\xi .$$

The norms in Schauder spaces $C^{(k),\mu}(\langle 0,1\rangle)$ will be denoted by $\|u\|^{k,\mu}$.

<u>Lemma 1.1.</u> If λ is an eigenvalue and u is the corresponding eigenvector, then

$$(1.9)\qquad \|u\|^{1,1/(p-1)} \leq c .$$

<u>Proof.</u> We have $df(u,v) - \lambda dg(u,v) = 0$, for each $v \in \overset{o}{W}{}_p^{(1)}$, hence

$$0 = \int_0^1 (a|u'|^{p-2} u'v' + (b - \lambda c)|u|^{p-2} uv)\, dx =$$

$$= \int_0^1 \Big[a(x)|u'(x)|^{p-2} u'(x) +$$

$$+ \int_{x_0}^{x} (\lambda c(\xi) - b(\xi))|u(\xi)|^{p-2} u(\xi)\, d\xi\Big] v'(x)\, dx =$$

$$= \int_0^1 M_{x_0}(x) v'(x)\, dx .$$

If $g \in L_p$, then

$$v(x) \overset{df}{=} \int_0^x (g(\xi) - \int_0^1 g(\eta)\, d\eta)\, d\xi \in \overset{o}{W}{}_p^{(1)} .$$

Put $c_1 = \int_0^1 M_{x_0}(x)\, dx$. Then for each $g \in L_p$, $0 = \int_0^1 (M_{x_0}(x) - c_1)g(x)\, dx$, hence

(1.10) $$M_{x_0}(x) = c_1 \quad \text{almost everywhere,}$$

which implies (1.9), q. e. d.

Lemma 1.2. If u is an eigenvector, then

(1.11) $$|u(x)| + |u'(x)| > 0 ,$$

$$[a(x)|u'(x)|^{p-2} u'(x)]'_{x=x_0} \neq 0 \quad \text{for} \quad u'(x_0) = 0 .$$

Proof. Let us suppose $|u(x_0)| + |u'(x_0)| = 0$. Then

(1.12) $$M_{x_0}(x) = 0$$

and

(1.13) $$|u(x)| \leq c_1|x - x_0| .$$

Substituting (1.13) into (1.12), we obtain

$$|u'(x)| \leq c_1 c_3 p^{-1/(p-1)} |x - x_0|^{p/(p-1)} \leq c_1 c_3 |x - x_0| ,$$

therefore $|u(x)| \leq c_1 c_3 |x - x_0|^2$.

Let us suppose $|u(x)| \leq c_1 c_3^k |x - x_0|^{1+k}$, k integer. Substituting again into (1.12), we obtain

$$|u'(x)| \leq c_1 c_3^{k+1} |x - x_0|^{\frac{(1+k)(p-1)+1}{p-1}} [(1 + k)(p - 1) + 1]^{-\frac{1}{p-1}} \leq$$

$$\leq c_1 c_3^{k+1} |x - x_0|^{1+k} ,$$

hence

$$(1.14) \qquad |u(x)| \leq c_1 c_3^{1+k} |x - x_0|^{2+k} .$$

If $|x - x_0| \leq \delta$ and $\delta c_3 < 1$, then it follows from (1.14) that $u(x) = 0$ in $\langle 0,1\rangle \cap \langle x_0 - \delta, x_0 + \delta\rangle$ and therefore $u(x) \equiv 0$ in $\langle 0,1\rangle$ which is impossible, q. e. d.

For a fixed eigenvector, put $\varrho(x) = |u'(x)|^{p-2}$ and let

$$W_{2,\varrho}^{(1)} = \left\{ h \mid \int_0^1 ((h')^2 + h^2)\varrho \, dx \overset{df}{=} \|h\|_{1,2,\varrho}^2 < \infty \right\} .$$

Lemma 1.3. If $1 \leq q < 2(p - 1)/(2p - 3)$, then

$$(1.15) \qquad W_{2,\varrho}^{(1)} \subset W_q^{(1)}$$

and the identity mapping is continuous. The linear hull of $\sin n\pi x$, $n = 1, 2, \ldots$ is dense in $\overset{o}{W}{}_{2,\varrho}^{(1)}$ $(\overset{o}{W}{}_{2,\varrho}^{(1)} = \{h \mid h \in W_{2,\varrho}^{(1)}$, $h(0) = h(1) = 0\})$.

Proof. (1.15) is a consequence of the Hölder inequality since it follows from Lemma 1.2 that $|u'(x)| \geq c|x - x_0|^{1/(p-1)}$ in the neighbourhood of a root x_0 of $u'(x)$.

$\int_0^1 u'v'\varrho \, dx$ is an innerproduct in $\overset{o}{W}{}_{2,\varrho}^{(1)}$. If

$$\int_0^1 u' \cos n\pi x \, \varrho \, dx = 0 , \quad n = 1, 2, \ldots,$$

then $u'\varrho = c$ and supposing $c > 0$ it is $u' > 0$, possibly with the exception of the finite number of roots of $u'(x)$. This is impossible because $0 = u(1) - u(0) = \int_0^1 u'(x) \, dx$.

Let us denote by V_ϱ the functions from $W_{2,\varrho}^{(1)}$ satisfying one of the conditions (1.1).

Lemma 1.4. Let for a fixed eigenvector u corresponding to λ, the following eigenvalue problem is considered:

$$d^2 f(u,v,h) - \mu d^2 g(u,v,h) = 0$$

for each $h \in V_\varrho$, $v \in V_\varrho$, $v \not\equiv 0$, $\mu \in R_1$. Then μ is a simple eigenvalue and $|u'|^{p-2} h'$ is once continuously differentiable.

Proof. Put

$$N_{x_0}(x) \overset{df}{=} a(x)|u'(x)|^{p-2} v'(x) +$$

$$+ \int_{x_0}^{x} (\mu c(\xi) - b(\xi))|u(\xi)|^{p-2} v(\xi)\, d\xi .$$

As in the proof of Lemma 1.1, $N_{x_0}(x) = \int_0^1 N_{x_0}(\xi)\, d\xi$ almost everywhere, hence $|u'(x)|^{p-2} v'(x)$ has one continuous derivative.

First, let us consider a space V_ϱ for which $w \in V_\varrho \Rightarrow w(0) = 0$ (the same for $w(1) = 0$). We have $u(0) = 0$ and by Lemma 1.2 $u'(0) \neq 0$, hence $v'(x)$ is continuous near the origin. Let us suppose that μ is not simple. Then there exists a nontrivial eigenvector v such that $v(0) = v'(0) = 0$. Hence $N_0(x) = 0$ and v satisfies the linear Volterra's integral equation

$$v(x) + \int_0^x \left(\int_\eta^x \frac{d\xi}{a(\xi)|u'(\xi)|^{p-2}} \right) (\mu c(\eta) -$$

$$- b(\eta))|u(\eta)|^{p-2} v(\eta)\, d\eta = 0$$

with a continuous kernel. Therefore $v = 0$, which is impossible.

It remains to consider the case (i) from (1.1).

From $df(u,h) - \lambda dg(u,h) = 0$ for each $h \in W_2^{(1)}$, we obtain

$$(1.16) \qquad (A_0 - \lambda B_0)|u(0)|^{p-2} u(0) = a(0)|u'(0)|^{p-2} u'(0) ,$$

$$(A_1 - \lambda B_1)|u(1)|^{p-2} u(1) = -a(1)|u'(1)|^{p-2} u'(1) .$$

If $B_0 = B_1 = 0$ and $A_0 > 0$ or $\lambda B_0 - A_0 > 0$ (the other possibilities are left to the reader), then $u'(0) \neq 0$. From $d^2f(u,v,h) - \mu d^2 g(u,v,h) = 0$ for each $h \in W_{2,\varrho}^{(1)}$ we obtain

$$(1.17) \qquad a(0)|u'(0)|^{p-2} v'(0) = (A_0 - \mu B_0)|u(0)|^{p-2} v(0) .$$

If μ is not simple, there exists a nontrivial eigenfunction such that $v(0) = 0$ and from (1.17) it follows that $v'(0) = 0$; as above, we come to a contradiction, q. e. d.

Lemma 1.5. If u_1 and u_2 are two eigenvectors such that

$$\|u_1\|_{1,p} = \|u_2\|_{1,p} = 1 \text{ and } \|u_1 - u_2\|_{1,p} \leq \frac{1}{2} ,$$

then

$$|\lambda_1 - \lambda_2| \leq c\|u_1 - u_2\|_{1,2}^2 .$$

Proof. λ and u are respectively an eigenvalue and an eigenvector iff for the Rayleigh's quotient

$$\Phi(u) \overset{df}{=} \frac{\int_0^1 (a|u'|^p + b|u|^p)\,dx + A_0|u(0)|^p + A_1|u(1)|^p}{\int_0^1 c|u|^p\,dx + B_0|u(0)|^p + B_1|u(1)|^p} ,$$

$\Phi(u) = \lambda$ and $d\Phi(u,h) = 0$ for each $h \in V$. Φ being twice continuously Frechet differentiable,

$$\lambda_2 - \lambda_1 = \int_0^1 \int_0^1 d^2\Phi(u_1 + t\tau(u_2 - u_1), u_2 - u_1, t(u_2 - u_1))\,dt\,d\tau .$$

The assertion follows in virtue of Lemma 1.1.

Proof of Theorem 1.1. First we prove the easier part: if the eigenvectors are isolated, then the set of all eigenvalues is isolated and to every λ_i it corresponds a finite number of normed eigenvectors. In fact, if $\lambda_{n_k} \to \lambda$, then we can suppose $u_{n_k} \to u$. As $f'(u_{n_k}) - \lambda_{n_k} g'(u_{n_k}) = 0$, g' is strongly continuous and f' satisfies condition (S), we obtain $(f'(u_{n_k}) - f'(u), u_{n_k} - u) \to 0$, hence $u_{n_k} \to u$ and therefore u is an eigenvector corresponding to λ, which is impossible because u is isolated. Further, we prove analogously as above that the set of eigenvectors corresponding to the same eigenvalue is compact; being isolated, it is necessarily finite.

Let us suppose that there exist u_n such that $u_n \to u$; u is therefore an eigenvector and $\lambda_n \to \lambda$.

Let us denote by $0 \leq x_1 < x_2 < \dots < x_t \leq 1$ the roots of $u'(x)$. It follows from Lemma 1.1 that we can suppose $u_n \to u$ in $C^{(1),\varkappa}$, $0 < \varkappa < 1/(p-1)$. Lemma 1.2 implies for $n \geq n_0$ that u_n' have the same number of roots as u' and that $x_i^n \to x_i$. Put $1 < q_1 < q_2 < 2(p-1)/(2p-3)$. We suppose $\|u_n\|_{1,q_1} = \|u\|_{1,q_1}$ and obtain

$$(1.18) \qquad \int_0^1 d^2 f(u + t(u_n - u), u_n - u, v)\,dt -$$

$$- \lambda \int_0^1 d^2 g(u + t(u_n - u), u_n - u, v)\, dt =$$

$$= (\lambda_n - \lambda)\, dg(u_n, v) \ , \ v \in V .$$

Put $v = u_n - u$ and divide (1.18) by $\|u_n - u\|^2_{1,q_1}$. Put

$$h_n = \frac{u_n - u}{\|u_n - u\|_{1,q_1}} .$$

First, consider $|u'(x) + t(u_n'(x) - u'(x))|^{p-2}$. For the sake of simplicity let us assume that the function $u'(x)$ has one root x_1 in the interval $\langle 0,1 \rangle$. There exists $\delta > 0$ such that

$$(1.19) \qquad |u'(x) + t(u_n'(x) - u'(x))|^{p-2} \geqq c_1 > 0$$

holds for $x \in \langle 0,1 \rangle \setminus (x_1 - \delta, x_1 + \delta)$, $t \in \langle 0,1 \rangle$, n large enough, and considering $-u$ instead of u if necessary,

$$(1.20) \qquad u'(x) \leqq -c_2 |x - x_1|^{\frac{1}{p-1}} , \ x_1 - \delta \leqq x \leqq x_1 ,$$

$$u'(x) \geqq c_3 |x - x_1|^{\frac{1}{p-1}} , \ x_1 \leqq x \leqq x_1 + \delta ,$$

$$(1.21) \qquad u_n'(x) \leqq -c_2 |x - x_1^n|^{\frac{1}{p-1}} , \ x_1 - \delta \leqq x \leqq x_1^n ,$$

$$u_n'(x) \geqq c_3 |x - x_1^n|^{\frac{1}{p-1}} , \ x_1^n \leqq x \leqq x_1 + \delta .$$

Let us suppose $x_1 < x_1^n$. Inequalities (1.20) and (1.21) imply for $x_1 - \delta \leqq x \leqq (x_1 + x_1^n)/2$ and $\alpha \leqq t \leqq 1$ with $0 < \alpha < 1$ the inequality

$$(1.22) \qquad |u'(x) + t(u_n'(x) - u'(x))|^{p-2} \geqq c_4 |x - x_1^n|^{\frac{p-2}{p-1}}$$

and for $(x_1 + x_1^n)/2 \leqq x \leqq x_1 + \delta$, $0 \leqq t \leqq \beta$, $0 < \beta \leqq 1$ the inequality

$$(1.23) \qquad |u'(x) + t(u_n'(x) - u'(x))|^{p-2} \geqq c_4 |x - x_1|^{\frac{p-2}{p-1}} .$$

Hence for $h \in W_2^{(1)}$ we have

$$\int_0^1 dt \int_0^1 |u'(x) + t(u_n'(x) - u'(x))|^{p-2} (h'(x))^2 \, dx \geq$$

$$\geq \int_\alpha^1 dt \int_0^{(x_1+x_1^n)/2} |u'(x) + t(u_n'(x) - u'(x))|^{p-2} (h'(x))^2 \, dx +$$

$$+ \int_0^\beta dt \int_{(x_1+x_1^n)/2}^1 |u'(x) + t(u_n'(x) - u'(x))|^{p-2} (h'(x))^2 \, dx \geq$$

$$\geq c_5\left(\int_0^1 |h'|^{q_2} \, dx\right)^{\frac{2}{q_2}} .$$

We arrive at the main inequality: for $h \in W_2^{(1)}$,

$$\int_0^1 d^2f(u + t(u_n - u),h,h) \, dt \geq c_5\left(\int_0^1 |h'|^{q_2} \, dx\right)^{\frac{2}{q_2}} . \tag{1.24}$$

It follows from Lemma 1.1 and 1.5 that

$$|\lambda - \lambda_n| \leq c_6 \|u - u_n\|_{1,q_1}^{q_1} \tag{1.25}$$

and hence finally from (1.18), (1.24) and (1.25)

$$\|h_n\|_{1,q_2}^2 \leq c_7\|h_n\|_{1,q_1}^2 + c_8\|u - u_n\|_{1,q_1}^{q_1-1} . \tag{1.26}$$

We can suppose $h_n \to h$ in $W_{q_2}^{(1)}$. Putting

$$\mathcal{O}(x) \overset{df}{=} a(x) \left(\int_0^1 |u'(x) + t(u_n'(x) - u'(x))|^{p-2} \, dt\right) h_n'(x) +$$

$$+ \int_0^x \left(\int_0^1 |u(\xi) + t(u_n(\xi) - u(\xi))|^{p-2} \, dt\right)(\lambda c(\xi) - b(\xi))h_n(\xi) \, d\xi +$$

$$+ \frac{\lambda_n - \lambda}{\|u_n - u\|_{1,q_1}} \int_0^x c(\xi) |u_n(\xi)|^{p-2} u_n(\xi) \, d\xi ,$$

we obtain as above

$$\mathcal{O}(x) = \int_0^1 \mathcal{O}(\xi) \, d\xi , \tag{1.27}$$

hence

$$\left(\int_0^1 |u'(x) + t(u_n'(x) - u'(x))|^{p-2} \, dt\right) h_n'(x) \overset{df}{=} \varphi_n(x)$$

has one continuous derivative and $\|\varphi_n'\|^{1,1} \le c$. Therefore we can suppose $h_n'(x) \to h'(x)$ everywhere, with the exception of the points $x_1, \dots, x_t$. The Vitali theorem implies $h_n \to h$ in $W_{q_1}^{(1)}$, therefore $\|h\|_{1,q_1} = 1$. The Fatou lemma gives, starting from (1.18): $h \in W_{2,\varrho}^{(1)}$ and therefore $h \in V_\varrho$. Taking the functions $\sin n\pi x + ax + b$ (with a , b chosen to satisfy condition (1.1)) in the space of v , we obtain finally

$$d^2 f(u,h,v) - \lambda d^2 g(u,h,v) = 0 \quad \text{for each} \quad v \in V_\varrho . \tag{1.28}$$

Clearly u is also an eigenfunction of (1.28) corrsponding to λ . Putting $\psi(u) \overset{df}{=} \|u\|_{1,q_1}^q$, we have

$$0 = \psi(u_n) - \psi(u) = d\psi(u,u_n - u) + \omega(u_n - u) ,$$

hence $d\psi(u,h) = 0$. On the other hand, $d\psi(u,u) \neq 0$, hence u is linearly independent of h , which ia a contradiction to Lemma 1.4, q. e. d.

§ 2. Operator of order 4

Let us consider

$$f(y) \overset{df}{=} \frac{1}{p}\int_0^1 a(x)|y''(x)|^p\,dx + \frac{1}{p}\int_0^1 b(x)|y(x)|^p\,dx , \tag{2.1}$$

$$g(y) \overset{df}{=} \frac{1}{p}\int_0^1 c(x)|y(x)|^p\,dx , \tag{2.2}$$

$V = \overset{o}{W}{}_p^{(2)}$. If λ_1 is the least eigenvalue, we suppose

$$\lambda_1 c(x) - b(x) > 0 . \tag{2.3}$$

As before, we suppose $a \in C^{(1)}$, $a(x) > 0$, b , $c \in C$, $b(x) \ge 0$, $c(x) > 0$.

Let us remark in the beginning of this section that some steps in the proofs of the lemmas and theorem below were inspired by a paper of S. A. Janczewsky [22].

<u>Theorem 2.1.</u> The eigenfunctions of the problem $f'(u) - \lambda g'(u) = 0$ for the functionals (2.1), (2.2) and the space $\overset{o}{W}{}_p^{(2)}$ are iso-

lated. Eigenvalues form a countable set $0 < \lambda_1 < \lambda_2 < \lambda_3 < \dots$ and $\lim_{n\to\infty} \lambda_n = \infty$. To every eigenvalue there corresponds a finite set of normed eigenfunctions such that with every u it contains also $-u$.

Proof. The main idea of the proof is the same as that of the proof of Theorem 1.1. That is why we leave those parts of the proof which are completely analogous to the reader. A complete proof can be found in the paper by A. Kratochvíl, J. Nečas [17]. We shall mention only those steps of the proof which differ essentially from the proof of Theorem 1.1.

For eigenvectors such that $\|u\|_{\overset{\circ}{2},p} \leq 2$ and eigenvalues such that $\lambda \leq N$ we obtain

$$\|u\|^{2,1/(p-1)} \leq c < \infty . \tag{2.4}$$

Lemma 2.1. If u is an eigenvector, $u''(x_0) = 0$, then

$$[a(x)|u''(x)|^{p-2} u''(x)]'_{x=x_0} \neq 0 .$$

Proof. Let us suppose the contrary. Then

$$a(x)|u''(x)|^{p-2} u''(x) = \tag{2.5}$$

$$= \int_{x_0}^{x} (x - \xi)(\lambda c(\xi) - b(\xi))|u(\xi)|^{p-2} u(\xi)\, d\xi .$$

If $u(x_0) \geq 0$ and $u'(x_0) > 0$, then it follows from (2.5) that there exists $\varepsilon > 0$ such that $u''(x) > 0$ for $x_0 < x \leq x_0 + \varepsilon$. Hence $u'(x)$ is increasing and therefore $u(x)$ is increasing. That is why we can start once more from the point $x_0 + \varepsilon$. Hence the set of $a > x_0$ for which $u(a) > 0$ is closed and open, therefore $u(1) > 0$, which is impossible. If $u(x_0) < 0$, $u'(x_0) > 0$, then $u''(x)$ is increasing in an interval $\langle -\varepsilon + x_0, x_0 \rangle$, hence $u''(x) < 0$ in $\langle -\varepsilon + x_0, x_0)$, therefore $u'(x)$ is increasing in $(-\varepsilon + x_0, x_0\rangle$ and so is $u(x)$. As before, we can once more start at the point $x_0 - \varepsilon$ and finally we obtain $u(0) < 0$, which is also impossible.

The only case which cannot be reduced to the preceding ones is $u(x_0) = u'(x_0) = 0$.

This is just the situation as in Lemma 1.2 and we obtain a contradiction in the same way as there.

Lemma 2.2. If u is an eigenvector, then

$$u''(0) \neq 0 \neq u''(1) .$$

Proof. Let us suppose $u''(0) = 0$. Then

$$a(x)|u''(x)|^{p-2} u''(x) = \tag{2.6}$$

$$= \int_0^x (x - \xi)(\lambda c(\xi) - b(\xi))|u(\xi)|^{p-2} u(\xi)\, d\xi + cx .$$

If $c = 0$, then $u(0) = u'(0) = 0$ implies $u(x) \equiv 0$ as above, which is impossible. If $c > 0$, we obtain from (2.6)

$$[a(x)|u''(x)|^{p-2} u''(x)]_{x=0}^{1} > 0 ,$$

hence $u''(x)$ is increasing in an interval $\langle 0,\varepsilon\rangle$ and therefore the same is true for $u'(x)$ and $u(x)$. As before, we obtain $u(1) > 0$, which is impossible, q. e. d.

Put $\varrho(x) = |u''(x)|^{p-2}$ and

$$V_\varrho = \left\{u \mid \int_0^1 ((u'')^2 + (u')^2 + u^2)\varrho\, dx \overset{df}{=} \|u\|^2_{2,2,\varrho} < \infty ,\right.$$

$$\left. u(0) = u'(0) = u(1) = u'(1) = 0\right\} .$$

Lemma 2.3. If u is an eigenfunction, then the eigenvalues μ of the equation in variations

$$d^2f(u,v,h) - \mu d^2g(u,v,h) = 0 \quad \text{for each} \quad h \in V_\varrho$$

are simple.

Proof. As $u''(0) \neq 0$, $v \in C^{(2)}$ in a neighbourhood of the origin. If μ is not simple, then there exists an eigenvector v such that $v(0) = v'(0) = v''(0) = 0$.

We have

$$a(x)|u''(x)|^{p-2} v''(x) =$$

$$= \int_0^x (x - \xi)(\lambda c(\xi) - b(\xi))|u(\xi)|^{p-2} v(\xi)\, d\xi + cx .$$

If $c = 0$, we obtain as in Lemma 1.4 a homogeneous Volterra integral equation with a continuous kernel, hence $v \equiv 0$. The case $c > 0$ leads to $v(1) > 0$ as before, which is a contradiction.

All other steps of the proof of Theorem 2.1 are completely analogous to those of the proof of Theorem 1.1.

BIBLIOGRAPHY

[1] S. I. Pochožajev: On the Solvability of Nonlinear Equations Involving Odd Operators. Funct. Anal. and Appl. (Russian) 1 (1967), 66 - 73.

[2] J. Nečas: Sur l'alternative de Fredholm pour les opérateurs non-linéaires avec applications aux problèmes aux limites. Annali Scuola Normale Sup. Pisa Vol. XXIII, Fasc. II (1969), 331 - 345.

[3] J. Nečas: Fredholm Alternative for Nonlinear Operators and Applications to Partial Differential Equations and Integral Equations. To appear in Časopis pro pěstování matematiky.

[4] F. E. Browder: Existence Theorems for Nonlinear Partial Differential Equations. Proc. Amer. Math. Soc. 1968 Summer Institute in Global Analysis (to appear).

[5] M. Kučera: Fredholm Alternative for Nonlinear Operators. Comm. Math. Univ. Carol. 11, 2 (1970), 337 - 363.

[6] S. Fučík: Note on the Fredholm Alternative for Nonlinear Operators. Comm. Math. Univ. Carol. 12 (1971), 213 - 226.

[7] R. I. Kačurovskij: On Fredholm Theory for Nonlinear Operator Equations. Dokl. Akad. Nauk SSSR 192 (1970), No. 5 (Russian).

[8] E. S. Citlanadze: Theorems of the Existence of Minimax Points in Banach Spaces (Russian). Trudy Mosk. Mat. Obšč. 2 (1953), 235 - 274.

[9] S. Fučík, J. Nečas: Ljusternik-Schnirelmann Theorem and Nonlinear Eigenvalue Problem. To appear in Mathematische Nachrichten.

[10] J. Naumann: Ljusternik-Schnirelmann Theorie und nichtlineare Eigenwertprobleme. To appear in Mathematische Nachrichten.

[11] M. M. Vajnberg: Variational methods for the study of nonlinear operators. Holden - Day 1964 (Russian: Moscow 1956).

[12] M. A. Krasnoselskij: Topological methods in the theory of nonlinear integral equations. Pergamon Press 1964 (Russian: Moscow 1956).

[13] M. S. Berger, M. S. Berger: Perspectives in nonlinearity. W. A. Benjamin 1968.

[14] J. T. Schwartz: Nonlinear functional analysis. Courant Institute, New York 1965.

[15] S. I. Pochožajev: The Set of Critical Values of a Functional. (Russian). Mat. Sbornik 75 (1968) (117).

[16] J. Nečas: On the Discreteness of the Spectrum of Nonlinear Sturm-Liouville Equation (Russian). Dokl. Akad. Nauk SSSR.

[17] A. Kratochvíl, J. Nečas: On the Discreteness of the Spectrum to Nonlinear Sturm-Liouville Equation of the 4th Order (Russian). Comm. Math. Univ. Carol. 12 (1971), 639 - 653.

[18] I. Kupka: Counterexample to Morse-Sard Theorem in the Case of Infinite Dimensional Manifolds. Proc. Amer. Math. Soc. 16 (1965), 954 - 957.

[19] A. Marino, S. Spagnolo: Una nota sul lemma de Sard in infinite dimensioni. Boll. Unione Mat. Ital. (1968 or 1969).

[20] S. Fučík, J. Milota: On the Convergence of Sequences of Linear Operators and Adjoint Operators. Comm. Math. Univ. Carol. 12 (1971), 753 - 763.

[21] S. Fučík, O. John, J. Nečas: Schauder Bases in Sobolev Spaces. To appear in Comm. Math. Univ. Carol.

[22] S. A. Janczewsky: Oscillation Theorems for the Differential Boundary Value Problems of the Fourth Order.I: Ann. of Math. 29 (1928), 521 - 542; II: Ann. of Math. 31 (1930), 663 - 680.

[23] S. L. Troyanski: On Locally Uniformly Convex and Differentiable Norms in Certain Non-separable Banach Spaces. Studia Math. 37 (1971), 173 - 180.

[24] S. Fučík, J. Nečas, J. Souček, V. Souček: Upper Bound for the Number of Eigenvalues for Nonlinear Operators. Ann. Scuola Norm. Sup. Pisa (1972), to appear.

[25] S. Fučík, J. Nečas, J. Souček, V. Souček: Upper Bound for the Number of Critical Levels for Nonlinear Operators in Hilbert Spaces of the Type of Second Order Nonlinear Partial Differential Operators. To appear.

[26] S. Fučík, J. Nečas, J. Souček, V. Souček: New Infinite Dimensional Versions of Morse-Sard Theorem. To appear.

Matematický ústav ČSAV, Žitná 25, Praha 1, Czechoslovakia

THE MORSE-SARD THEOREM FOR REAL-ANALYTIC FUNCTIONS

JIŘÍ SOUČEK, VLADIMÍR SOUČEK,
PRAHA (CZECHOSLOVAKIA)

Let us consider real functions $f(x) = f(x_1,\dots,x_N)$, defined on the domain $U \subset E_N$. It is well-known that

$$f \in C^k(U)\ ,\ k \geqq N \Rightarrow H_{\frac{N}{k}}(f(Z)) = 0$$

where $Z = \{x \in U\ ;\ \text{grad}\, f(x) = 0\}$ and $H_{\frac{N}{k}}$ is $\frac{N}{k}$-dimensional Hausdorff measure.

It is possible to prove also (see [2]) that

$$f \in C^{k+\alpha}(U)\ ,\ k \geqq N\ ,\ \alpha > 0 \Rightarrow H_{\frac{N}{k+\alpha}}(f(Z)) = 0\ .$$

Because it is easy to construct an innumerable set $M \subset \langle 0,1\rangle$ such that $H_\lambda(M) = 0\ ,\ \forall\lambda > 0$, there is question, which assumption must be made to obtain that the set $f(Z)$ is at most countable. It is natural then to consider real-analytic functions (i. e. the functions, which have a power series expansion in neighborhood of each point) and indeed, for such functions there holds even a little more:

Theorem. Let $D \subset R_N$ be an open set and let f be a real-analytic function defined on D. Denote

$$B = \{x \in D : \frac{\partial f(x)}{\partial x_i} = 0\ ,\ i = 1, \dots, N\}\ .$$

Then for each $x_0 \in B$ there exists a neighborhood $U(x_0) \subset D$ of the point x_0 such that $f(B \cap U(x_0))$ is a one-point set.

Moreover, for each compact set $K \subset D$ the set $f(B \cap K)$ is finite and the set $f(B)$ is countable.

The proof of the theorem is based on some theorems about germs of varieties from the theory of functions of several complex variables. We recapitulate for the reader necessary definitions and theorems from the book by R. C. Gunning, H. Rossi [1] (in brackets we refer to the numbers of definitions and theorems of this book).

An open polydisc in C_N is a subset $\Delta(w;r) \subset C_N$ of the form $\Delta(w;r) = \Delta(w_1,\dots,w_N;r_1,\dots,r_N) = \{z \in C_N : |z_j - w_j| < r_j , 1 \leq j \leq N\}$. A complex-valued function f defined on an open subset $D \subset C_N$ is called holomorphic in D if each point $w \in D$ has an open neighborhood U , $w \in U \subset D$ such that the function f has a power series expansion

$$f(z) = \sum_{\nu_1 \dots \nu_N = 0}^{\infty} a_{\nu_1 \dots \nu_N} (z_1 - w_1)^{\nu_1} \dots (z_N - w_N)^{\nu_N}$$

which converges for all $z \in U$.

In following we will consider the polydiscs $\Delta(w;r)$ with $w = 0$.

Let X , Y be subsets of C_N . The sets X and Y are said to be equivalent at 0 if there is a neighborhood U of 0 such that $X \cap U = Y \cap U$. An equivalence class of sets is called the germ of a set. The equivalence class of X is to be denoted by $\mathcal{X}$ (see II.H. 4).

Each set $Y \in \mathcal{X}$ is called the representative of the germ $\mathcal{X}$. We define germ $\mathcal{V}_1 \cup \dots \cup \mathcal{V}_k$ as the germ the representative of which is the set $V_1 \cup \dots \cup V_k$, where V_i is a representative of $\mathcal{V}_i$ $(i = 1, \dots, k)$. Similarly, we say that $\mathcal{V}_1 \subset \mathcal{V}_2$ if for all representatives $V_1 \in \mathcal{V}_1$, $V_2 \in \mathcal{V}_2$ there exists a polydisc $\Delta = \Delta(0;r)$ such that $V_1 \cap \Delta \subset V_2 \cap \Delta$. It can be easily seen that these definitions are correct.

A germ $\mathcal{V}$ is the germ of a variety if there are a polydisc $\Delta = \Delta(0;r)$ and functions $f_1, \dots, f_t$ holomorphic in Δ such that

$$\{x \in \Delta : f_i(x) = 0 , i = 1, \dots, t\}$$

is a representative of $\mathcal{V}$ (see II.E.6).

We shall denote the collection of germs of a variety at the point 0 by $\mathcal{B}_N$. It can be proved, that $\mathcal{V}_1 \cup \mathcal{V}_2 \in \mathcal{B}_N$ for each $\mathcal{V}_1$, $\mathcal{V}_2 \in \mathcal{B}_N$ (see II.E.7).

A germ $\mathcal{V} \in \mathcal{B}_N$ is said to be irreducible if $\mathcal{V} = \mathcal{V}_1 \cup \mathcal{V}_2$ for $\mathcal{V}_1$, $\mathcal{V}_2 \in \mathcal{B}_N$ implies that either $\mathcal{V} = \mathcal{V}_1$ or $\mathcal{V} = \mathcal{V}_2$, i. e., the irreducible germ cannot be written as the nontrivial union of two germs from $\mathcal{B}_N$ (see II.E.12).

Theorem A (II.E.15). Let $V \in \mathcal{O}_N$. Then there exist irreducible germs $V_1, \ldots, V_k \in \mathcal{O}_N$ such that $V = V_1 \cup \ldots \cup V_k$.

A subset M of C_N is said to be a complex submanifold of C_N if to every point $p \in M$ corresponds a neighborhood U of the point p, a polydisc $\Delta = \Delta(0;r)$ in C_K ($K \leq N$) and a nonsingular holomorphic mapping $F : \Delta \to C_N$ such that $F(0) = p$, and $M \cap U = F(\Delta)$ (see I.B.8, I.B.10).

Theorem B. Let $V \in \mathcal{O}_N$ be an irreducible germ. Then there exist a polydisc $\Delta = \Delta(0;r)$ and a set $V \subset \Delta$ such that:

(i) $\overline{V}$ is a representative of V,

(ii) for each polydisc $\Delta_1 \subset \Delta$ there exists a polydisc $\Delta_2 \subset \Delta_1$ such that $V \cap \Delta_2$ is a connected complex submanifold.

This theorem follows immediately from III.A.10, III.A.9 and III.A.8, this is only reformulation of a part of Theorem III.A.10. First we shall prove the following easy lemma:

Lemma. Let f be a holomorphic function defined on a polydisc $\Delta = \Delta(0;r)$ and let $V \subset \Delta$ be a connected complex submanifold. Suppose that

$$(*) \qquad \frac{\partial f(z)}{\partial z_i} = 0$$

for all $z \in V$ and $i = 1, \ldots, N$.

Then the function f is constant on $\overline{V} \cap \Delta$, i. e., $f(\overline{V} \cap \Delta)$ is a one-point set.

Proof. It is sufficient to prove that f is constant on V, for f is continuous. Let $z_0 \in V$ be fixed. Let us denote

$$M = \{z \in V : f(z) = f(z_0)\} .$$

The set M is closed with respect to V, for f is continuous. We shall prove that M is open with respect to V. Suppose $z_1 \in M \subset V$. From the definition of complex submanifold it follows that there exist a neighborhood $U \subset C_N$ of the point z_1, polydisc $\Delta_1 \subset C_K$ ($K \leq N$) and a holomorphic mapping $F : \Delta_1 \to C_N$ such that

$$F(0) = z_1 \quad \text{and} \quad F(\Delta_1) = V \cap U .$$

Let $z \in V \cap U$ be fixed. Then there exists $\xi \in \Delta_1$ such that $F(\xi) = z$. Let us denote

$$\gamma(t) = F(t\xi) \quad \text{for} \quad t \in \langle 0,1\rangle .$$

Clearly $\gamma = (\gamma_1, \ldots, \gamma_N) : \langle 0,1\rangle \to V \cap U$ is a differentiable curve and

$$\frac{d}{dt}[f(\gamma(t))] = \sum_{i=1}^{N} \frac{\partial f(\gamma(t))}{\partial z_i} \cdot \frac{d\gamma_i(t)}{dt} = 0$$

by the assumption (*). Thus we have

$$f(z) - f(z_1) = f(\gamma(1)) - f(\gamma(0)) = 0$$

hence $V \cap U \subset M$.

Now, V is connected, M is open and closed with respect to V, hence $V = M$.

Proof of the theorem. Without loss of generality we can suppose $x_0 = 0$. Then we can write

$$f(x_1, \ldots, x_N) = \sum_{\nu_i \geq 0} a_{\nu_1 \ldots \nu_N} x_1^{\nu_1} \ldots x_N^{\nu_N} ,$$

where the above series converges uniformly for all $|x_1| < r_1, \ldots, |x_N| < r_N$. Now, it is possible to define (for $z \in \Delta(0;r) = \Delta_0$)

$$\tilde{f}(z_1, \ldots, z_N) = \sum_{\nu_i \geq 0} a_{\nu_1 \ldots \nu_N} z_1^{\nu_1} \ldots z_N^{\nu_N} .$$

We denote

$$V = \{z \in \Delta_0 : \frac{\partial \tilde{f}(z)}{\partial z_i} = 0 , \; i = 1, \ldots, N\} .$$

Clearly $f = \tilde{f}$ on $\Delta_0 \cap R_N$ and $B \cap \Delta_0 \subset V$. The germ $\mathcal{V}$ of the set V is by the definition the germ of variety, i. e., $\mathcal{V} \in \mathcal{B}_N$. From Theorem A it follows that there exist the irreducible germs $\mathcal{V}_1, \ldots, \mathcal{V}_k \in \mathcal{B}_N$ such that $\mathcal{V} = \mathcal{V}_1 \cup \ldots \cup \mathcal{V}_k$. For each germ $\mathcal{V}_i$ we can choose by Theorem B a polydisc Δ_1^i and the set V_i such that (i) and (ii) from Theorem B hold. Denote $\Delta_1 = \bigcap_{i=1}^{k} \Delta_1^i$. Further, there exists a polydisc $\Delta_2 \subset \Delta_1$ such that $V \cap \Delta_2 = (\overline{V}_1 \cap \Delta_2) \cup \ldots \ldots \cup (\overline{V}_k \cap \Delta_2)$, for $\overline{V}_i$ are representatives of $\mathcal{V}_i$. From Theorem B it follows that for every $i = 1, \ldots, k$ there exists a polydisc $\Delta_3^i \subset \Delta_2$ such that $V_i \cap \Delta_3^i$ is a connected complex submanifold. We have $\partial\tilde{f}(z)/\partial z_i = 0$ for all $z \in V_i \cap \Delta_3^i$, for $V_i \cap \Delta_3^i \subset V \cap \Delta_2 \subset \subset V$. By the previous lemma the function $\tilde{f}$ is constant on $\overline{V}_i \cap$

$\cap \Delta_3^i$. Denote $\Delta_3 = \bigcap_{i=1}^{k} \Delta_3^i$. We can choose $\Delta_4 \subset \Delta_3$ such that for every $i = 1, \ldots, k$ either $0 \in \overline{V}_i \cap \Delta_4$ or $\overline{V}_i \cap \Delta_4 = \emptyset$, for the sets $\overline{V}_i \cap \Delta_3$ are closed with respect to Δ_3 . Now, for every $z \in \overline{V}_i \cap \Delta_4$ there exists i , $1 \leqq i \leqq k$, such that $z \in \overline{V}_i \cap \Delta_4$, hence $\tilde{f}(z) = \tilde{f}(0)$. The set $\tilde{f}(V \cap \Delta_4)$ is then one-point set, and thus also the set $f(B \cap (\Delta_4 \cap R_N))$ is a one-point set.

The assertion about the set $f(B \cap K)$ follows immediately from the compactness of $B \cap K$. Further, we can choose the compact sets K_n such that $\bigcup_{n=1}^{\infty} K_n = D$, hence $f(B)$ is at most countable.

REFERENCES

[1] R. C. Gunning, H. Rossi: Analytic functions of several complex variables. Prentice Hall.

[2] M. Kučera: Hausdorff measures of the set of critical values. Comm. Math. Univ. Carolinae 13, 2 (1972).

[3] J. Souček, V. Souček: The Morse-Sard theorem for real-analytic functions. Comm. Math. Univ. Carolinae 13, 1 (1972).

J. Souček, Matematický ústav ČSAV, Žitná 25, Praha 1, Czechoslovakia
V. Souček, Matematicko-fyzikální fakulta KU, Sokolovská 83, Praha 8, Czechoslovakia

ON THE SOLVABILITY OF SOME NON-LINEAR EQUATIONS*)

SERGIO SPAGNOLO, PISA (ITALY)

We shall describe here some "abstract" theorems on global surjectivity (Theorems 1 and 2) and local surjectivity for a class of monotone operators, and an application of these results to the theory of differential equations (Theorem 4).

Theorems 1, 2 and also Theorem 3, except for its local nature, can be viewed in relation with the following well known theorem due to F. Browder of 1963 (see [1]): Every operator $A : V \to V'$ which acts between a real Banach space and its dual and which is demi-continuous, monotone and asymptotically coercive (i. e. it holds $\lim_{\|v\|_V \to \infty} \langle Av, v\rangle / \|v\|_V = +\infty$) is necessarily surjective.

The operators with which we shall be concerned here are again monotone, however, they are coercive with respect to a norm weaker than that of V. This weakened condition of coercivity will not entirely exclude the possibility of solving the equation $Au = f$; it will, however, be necessary to choose f in a suitable subspace Y of V' and to seek the solution u in a space $X \supset V$ (on which the above defined operator A should be considered). In other words, there may be a loss in the "regularization" of the solution with respect to the right hand side, either partly or completely (when $X \equiv V$).

A concrete example of operators of this type occurs in the theory of symmetric systems of positive type, introduced by Friedrichs ([2]) in the linear case and by Moser in the non-linear case. These systems in the linear case are of the form

*) The contents of this communication is taken from the paper "Sulla risolubilità di alcune equazioni non lineari" of A. Marino and the author, to be published elsewhere.

$$(1)\qquad L(u) \equiv b(x)u + \sum_{1}^{n} a_i(x) D_{x_i} u = f ,$$

with symmetric matrices $b(x)$ and $a_i(x)$ such that we have, for all x , ξ (denoting by I the identity matrix):

$$(2)\qquad b(x) - \frac{1}{2}\sum_{1}^{n} D_{x_i} a_i(x) \geq \gamma I ,$$

$$(3)\qquad k\sum_{1}^{n}{}_{i,j} D_{x_i} a_j(x)\xi_i\xi_j + \left[b(x) - \frac{1}{2}\sum_{1}^{n} D_{x_i} a_i(x)\right]|\xi|^2 \geq \gamma|\xi|^2 I ,$$

where γ and k are positive constants; in the non-linear case they are of the type

$$(4)\qquad A(u) \equiv F(x,u,D_{x_1}u,\ldots,D_{x_n}u) = f$$

where $F(x,y,p_1,\ldots,p_n)$ is such that

$$A'_w(u) \equiv F_y(x,w,Dw)u + \sum_{1}^{n} F_{p_i}(x,w,Dw)D_{x_i}u$$

is of type (1) with (2) and (3).

There are systems of degenerate type whose solution can be very slightly regular even when the right hand side is highly regular. A simple example ([3], page 293) shows in fact that there is a maximum regularity for the solution, which can be obtained in relation to the parameter k which occurs in (3), in the sense that there are solutions of (1), with f of class C^∞ , which belong to the Sobolev space H^k but not to any H^r with $r > k$.

Moser has shown in [3], combining the methods of Faedo-Galerkin and Newton, that if $F(x,y,p)$ satisfies the above assumptions at the origin (that is, if (2) and (3) hold with $b(x) = F_y(x,0,0)$, $a_i(x) = F_{p_i}(x,0,0)$) for some $k > k_0(n)$ and if, moreover, F is periodic in x with period 2π , is sufficiently regular and such that $F(x,0,0) \equiv 0$, then the equation (4) admits a unique solution of class C^2 for every sufficiently regular periodic f with period 2π , having sufficiently small sup norm.

Further, in the case of F linear in p (the quasi-linear case) the solution is shown to belong to H^k ; that is, it has the maximum possible regularity.

We are in a position to improve the result of Moser by making use of our Theorem 3 showing (Theorem 4) that the solution u belongs to H^k in every case.

Theorem 1. Let V be a real, reflexive and separable Banach space, V' its dual and let X be another Banach space in which is continuously densely immersed. Suppose further that $A : X \to V'$ is a demicontinuous operator (i. e., A transforms convergent sequences of X into weakly convergent sequences of V') such that

(i) $A(0) = 0$,

(ii) $\langle Av_1 - Av_2, v_1 - v_2 \rangle \geq 0$, $\forall\, v_1 , v_2 \in V$,

(iii) $\lim_{v \in V, \|v\|_X \to +\infty} \langle Av, v \rangle / \|v\|_X = +\infty$,

(iv) A transforms closed balls of X into closed subsets of V' .

Then, setting $X' = \{v' \in V' : |\langle v', v \rangle| \leq C(v')\|v\|_X\}$, we have $A(X) \supseteq X'$.

Remarks. Assumption (iv) is weaker than each one of the following assumptions, which hold in several concrete cases.

$(iv)_1$ A transforms weakly convergent sequences of X into weakly convergent sequences of V' (for example: A is linear).

$(iv)_2$ A is the restriction of a demi-continuous operator $\tilde{A} : \tilde{X} \to \tilde{V}'$ where $\tilde{X}$ and $\tilde{V}'$ are linear topological spaces such that X is compactly imbedded in $\tilde{X}$ and V' is continously imbedded in $\tilde{V}'$.

$(iv)_3$ There exists a linear and continuous duality which is separating in the second argument, $[\ ,\ ] : V' \times X \to R$ for which $[Ax_1 - Ax_2, x_1 - x_2] \geq 0$, $\forall\, x_1 , x_2 \in X$ (for example: V is a Hilbert space and A is monotone also with respect to the dual product on V'). [The fact that $(iv)_1$ and $(iv)_2$ imply (iv) is obvious; to show that $(iv)_3$ implies (iv) it is enough to use the artifice of Minty.]

Particular cases. (1) X is a Hilbert space, V is a Banach space which is continuously and densely imbedded in X , and X is considered to be imbedded in V' by transposition (so that $\langle x, v \rangle = (x, v)_X$, $\forall\, x \in X$, $v \in V$). Then, if A maps demi-continuously X into V' and V into X and further, if $(Av_1 - Av_2, v_1 - v_2)_X \geq \|v_1 - v_2\|_X^2$ and (iv) holds, then $A(X) \supseteq X$.

(2) $V \equiv X$: in this case we have $X' \equiv V'$ and hence (by choosing $\langle\ ,\ \rangle$ itself as the duality $[\ ,\ ]$, so that the assumption $(iv)_3$ holds) we obtain Browder's Theorem.

Theorem 2. Let X , Y be two real Hilbert spaces, X being continuously and densely imbedded in Y and let V be a linear subspace of X on which there exists a separating duality $\langle \, , \, \rangle : Y \times V \to R$ which is continuous in the first argument and coincides with the scalar product of X on $X \times V$. Suppose further that $A : X \to Y$ is a demi-continuous operator such that

(i) $$(Ax_1 - Ax_2, x_1 - x_2)_Y \geq C\|x_1 - x_2\|_Y^2 \, , \quad \forall \, x_1 \, , \, x_2 \in X \, ,$$

(ii) $$\langle Av, v\rangle \geq C\|v\|_X^2 - g(\|v\|_Y)\|v\|_X - h(\|v\|_Y) \, , \quad \forall \, v \in V \, ,$$

with $C > 0$ and continuous functions $g(t)$, $h(t)$. Then $A(X) \supseteq X$.

Theorem 3. Let X, Y and V be the same as in Theorem 2 and $B_r = \{x \in X \, ; \, \|x\|_X < r\}$. Suppose further that $A : B_\varrho \to Y$ is a Lipschitz continuous mapping (demi-continuous if the imbedding $X \to Y$ is compact) such that

(i) $$A(0) = 0 \, ,$$

(ii) $$(Ax_1 - Ax_2, x_1 - x_2)_Y \geq C_1\|x_1 - x_2\|_Y^2 \, , \quad \forall \, x_1 \, , \, x_2 \in B_\varrho \, ,$$

(iii) $$\langle Av, v\rangle \geq C_1\|v\|_X^2 - C_2\|v\|_Y^\vartheta\|v\|_X \, , \quad \forall \, v \in B_\varrho \cap V \, ,$$

where ϱ, ϑ, C_1 are constants > 0 and $C_2 \geq 0$. Then for every $0 < \varepsilon \leq \varrho$ there exists a $\delta = \delta(\varepsilon, \vartheta, C_1, C_2) > 0$ such that $A(B_\varepsilon) \supseteq B_\delta$.

Theorem 4. Let $F(x, y, p_1, \ldots, p_n)$ be a vector-valued function defined on the open subset $R_x^n \times \{|y| + |p_1| + \ldots + |p_n| < \sigma\}$ of $R_x^n \times R_y^m \times R_{p_1}^m \times \ldots \times R_{p_n}^m$ with values in R^m, 2π-periodic in x, of class C^{k+1} with $k > n/2 + 2$ and such that (denoting by I the identity matrix):

(a) $$F(x, 0, 0) = 0 \, ,$$

(b) $$b_0(x) \equiv F_y(x, 0, 0) - \frac{1}{2}\sum_1^n {}_i \, F_{x_i p_i}(x, 0, 0) \geq \gamma I > 0 \, ,$$

(c) $$k \sum_1^n {}_{i,j} \, F_{x_i p_j}(x, 0, 0)\xi_i\xi_j + b_0(x)|\xi|^2 \geq \gamma|\xi|^2 \, I \, .$$

Then there exists an $\varepsilon_0 = \varepsilon_0(\sigma, \gamma, k) > 0$ such that, for any $0 < \varepsilon \leq \varepsilon_0$, one can find a $\delta > 0$ in such a way that the system

$$F(x, u, D_{x_1}u, \ldots, D_{x_n}u) = f$$

has a (unique) solution u , 2π-periodic and with $\sum_{|\gamma|\leq k} \int_\Omega |D^\gamma u|^2\, dx < \varepsilon$, for every right hand side f which is 2π-periodic and satisfies $\sum_{|\gamma|\leq k} \int_\Omega |f|^2\, dx < \delta$ (we have denoted by Ω the period-parallelogram).

<u>Sketch of the proofs of the theorems. Theorem 1</u>. Since V is reflexive and separable, it can be renormed in such a way that there exists a demi-continuous duality mapping $J : V \to V'$ for which $\|Jv\|_{V'} = \|v\|_V$ and $\langle Jv, v\rangle = \|v\|_V^2$. One can then apply Browder's Theorem to the operators $A_\varepsilon = A + \varepsilon J$ and hence one can find, for every $y_0 \in V'$, a solution x_ε of $A_\varepsilon x_\varepsilon = y_0$. If now y_0 is chosen in X' then $\{x_\varepsilon\}$ is bounded in X and we can pass to the limit in the weak topology of X .

<u>Theorem 2</u>. Under the assumptions made on V , X , Y we see that V is a dense subspace of the Banach space Y_0 consisting of all $x \in X$ for which $\|x\|_{Y_0} \equiv \sup |(x,y)_X|/\|y\|_Y < +\infty$. We can still extend the assumption (ii) to every $v \in Y_0$ and hence there is no loss of generality in supposing that $V = Y_0$.

Now we prove that the spectral projections $\{p_n\}$ relative to the selfadjoint operator $ii^* : Y \to Y$ (where $i : X \to Y$ denotes the given imbedding and $i^* : Y \to X$ its adjoint) are selfadjoint with respect to both scalar products of X and Y . Hence it follows that, if we set $V_n = p_n(Y)$, the topologies of X and Y coincide on V_n . Applying then the theorem of Minty to the operators $p_n A/V_n : V_n \to V_n$, we can find, for every $y_0 \in Y$, the $x_n \in V_n$ for which we have $p_n A x_n = p_n y_0$; if now y_0 is chosen to be in X then we have $\|x_n\|_X \leq C$ and hence, passing to the limit in the weak topology of X , we obtain a solution x_0 of the equation $Ax_0 = y_0$.

<u>Theorem 3</u>. The projections p_n and the subspaces V_n are constructed as in the proof of Theorem 2 (we note that if the immersion of X in Y is compact then $\dim V_n < \infty$). On each V_n , we consider for $y_0 \in Y$ the Cauchy problem

$$(*) \qquad u'(t) = -p_n A(u(t)) + p_n y_0$$

with the initial condition $u(0) = 0$.

The hypothesis on A permits us to obtain the following "a priori" estimates for a solution $u(t)$ of $(*)$:

$$\|u(t+h) - u(t)\|_Y \leq e^{-C_1 t}\|u(h)\|_Y \quad \text{for} \quad h \geq 0 ,$$

$$\|u(t)\|_Y \leq \frac{1 - e^{-C_1 t}}{C_1} \|y_0\|_Y ,$$

$$\|u(t)\|_X \leq \frac{1 - e^{-C_1 t}}{C_1} (\|y_0\|_X + C_2 \|y_0\|_X^{\vartheta}) .$$

Using these estimates and the fact that on V_n the norms of X and Y are equivalent, we can find, for every y_0 in X with $\|y_0\|_X$ sufficiently small, a solution $u_n(t)$ of (•) defined on the whole of the half line $t \geq 0$ such that $x_n \equiv \lim_{t\to+\infty} u_n(t)$ exists and $\lim_{t\to+\infty} u_n'(t) = 0$, $\|x_n\|_X \leq \varrho/2$. But then we have $p_n A x_n = p_n y_0$, from which by the usual procedure we find an x_0 in X for which $Ax_0 = y_0$.

Theorem 4. Consider the spaces $V = [C_{2\pi}^{\infty}]^m$, $X = [H_{2\pi}^{k}]^m$ and $Y = [H_{2\pi}^{0}]^m$, where $C_{2\pi}^{\infty}$ is the space of all 2π-periodic C^{∞} functions on R^n, $H_{2\pi}^k$, for an integer $k \geq 0$, is the completion of $C_{2\pi}^{\infty}$ with respect to the scalar product $(u,v)_{H^k} =$
$= \sum_{|r|=k} C(r) \int_{\Omega} D^r u \, D^r v \, dx + \int_{\Omega} uv \, dx$, where $C(r) = k!/(r_1! \ldots \ldots r_n!)$, while for k real ≥ 0 one can define $H_{2\pi}^k$ by means of the coefficients in the Fourier development. The spaces V, X, Y satisfy the hypothesis of Theorem 2 with respect to the duality $\langle u,v\rangle = (-1)^k \sum_{|r|=k} C(r) \int_{\Omega} u D^{2r} v \, dx + \int_{\Omega} uv \, dx$, $u \in Y$, $v \in V$; moreover the immersion of X in Y is compact (Theorem of Rellich).

Assuming then $k > n/2 + 1$, it can be easily seen that the operator $A : u \mapsto F(x,u,D_1 u,\ldots,D_n u)$ is defined and continuous (in fact, of class C^1) on a ball B_{ϱ} with centre at the origin in X, with values in Y, and moreover, $A(0) = 0$, $A(B_{\varrho} \cap V) \subseteq X$.

We have therefore to prove that A satisfies assumptions (ii) and (iii) of Theorem 3 (restricting B_{ϱ} if necessary): Here (ii) follows easily from the assumption (b), by choosing $\varrho \leq \varrho_0(M,\gamma)$ where M is the supremum of all the derivatives of the function $F(x,y,p)$ of orders $\leq k+1$ on $|y| + \sum_1^n |p_i| \leq \sigma/2$.

Finally, to prove the condition (iii) of Theorem 3, we make use of the assumption (c) and the following two lemmas.

Lemma 1. Let $D^r = D_{i_1} \circ \ldots \circ D_{i_k}$, $k \geq 2$; then we have

$$D^r(Au) = F_{x_{i_1}\dots x_{i_k}}(x,u,Du) + F_y(x,u,Du)D^r u + \sum_1^n{}_i\, F_{p_i}(x,u,Du)D_i D^r u +$$

$$+ \sum_1^k{}_h \sum_1^n{}_i\, F_{x_{i_h}p_i}(x,u,Du)D_i D^{r(h)} u +$$

$$+ \sum_{|\alpha_1|=k,\,|\alpha_2|\leq 2} C_{r,\alpha_i} G_{r,\alpha_i}(x,u,Du)(D^{\alpha_1}u)(D^{\alpha_2}u) +$$

$$+ \sum_{\substack{\Sigma(|\alpha_i|-1)\leq k \\ |\alpha_i|\leq k-1 \\ h\leq k}} C_{r,\alpha_i} G_{r,\alpha_i}(x,u,Du)(D^{\alpha_1}u)\dots(D^{\alpha_h}u)$$

where $D^{r(h)}$ is such that $D^r = D_{i_h} \circ D^{r(h)}$; $G_{r,\alpha_i}(x,y,p)$ are suitable derivatives of orders $\leq k$ of $F(x,y,p)$ while C_{r,α_i} are constants.

Lemma 2. Let $|\alpha_i| \leq k - 1$, $\sum_1^h{}_i\, (|\alpha_i| - 1) \leq k$, $k > n/2 + 2$; then we have

$$\|(D^{\alpha_1}u)\dots(D^{\alpha_h}u)\|^{1/h}_{L^2(\Omega)} \leq C\|u\|^{1-\beta}_{H^k(\Omega)} \|u\|^{\beta}_{H^0(\Omega)}$$

with $\beta = \beta(k,\alpha_i) > 0$, $C = C(\Omega,k,\alpha_1,\dots,\alpha_h)$.

R E F E R E N C E S

[1] F. Browder. Bull. Amer. Math. Soc. 69 (1963), 862 - 874.

[2] K. O. Friedrichs. Comm. Pure Appl. Math. 11 (1958), 333 - 418.

[3] J. Moser. Ann. Sc. Norm. Pisa 20 (1966), 265 - 315.

Scuola Normale Superiore, 56100 Pisa, Italy

INTERPOLATION THEORY FOR SPACES OF BESOV TYPE. ELLIPTIC DIFFERENTIAL OPERATORS

HANS TRIEBEL, JENA (GDR)

CONTENTS

In the first section of this paper we give a summary of some results of the interpolation theory for spaces of Besov type proved in the last twelve years by Calderón, Grisvard, Lions, Magenes, Muramatu, Peetre, Taibleson, the author, and other mathematicians. We consider function spaces without weights and with weights defined in a domain Ω in the Euclidean n-space R_n including the Sobolev spaces W_p^k ; $k = 0, 1, \dots$; $1 < p < \infty$; the Slobodeckij spaces W_p^s ; $s > 0$; $s \neq$ integer ; $1 < p < \infty$; the Besov spaces $B_{p,q}^s$; $s > 0$; $1 < p < \infty$; $1 \leq q \leq \infty$; the Lebesgue spaces H_p^s ; $s > 0$; $1 < p < \infty$ and similar spaces with weights. For $\Omega = R_n$ we consider also spaces of distributions.

In the second section we describe some applications of the results of the first section to regular and singular elliptic differential operators.

We do not give any proofs of the results and refer the reader to the cited papers.

1. INTERPOLATION THEORY FOR SPACES OF BESOV TYPE

1.1. Abstract interpolation methods

We describe two interpolation methods, the "K-method" and the "complex method". The K-method is due to Peetre who showed in [42] the equivalence of this method to the other interpolation methods developed by Lions - Peetre in [34]. For a description of this method we refer also to [13]. The complex method is due to Lions, Calderón, and S. G. Krejn. A description of this method is given in [15], see also [26].

Let B_0 and B_1 be an interpolation couple of Banach spaces. That means that B_0 and B_1 are continuously embedded in a linear Hausdorff space. Then

$$B_0 + B_1 = \{u \mid u = u_0 + u_1 \ ; \ u_0 \in B_0 \ ; \ u_1 \in B_1\}$$

with the norm

$$K(t,u) = K(t,u,B_0,B_1) = \inf_{\substack{u=u_0+u_1 \\ u_0\in B_0,\, u_1\in B_1}} (\|u_0\|_{B_0} + t\|u_1\|_{B_1})$$

is a Banach space, t being a real number with $0 < t < \infty$. (All norms $K(t,u)$ are equivalent to each other.)

K-method [34, 42]. Let (B_0,B_1) be an interpolation couple. Let be $0 < \theta < 1$ and $1 \leq q \leq \infty$. We set

$$(B_0,B_1)_{\theta,q} = \Big\{u \mid u \in B_0 + B_1 \ ; \ \|u\|_{(B_0,B_1)_{\theta,q}} =$$

$$= \left[\int_0^\infty (t^{-\theta}K(t,u))^q \frac{dt}{t}\right]^{\frac{1}{q}} < \infty\Big\} ; \ 1 \leq q < \infty ,$$

$$= \{u \mid u \in B_0 + B_1 \ ; \ \|u\|_{(B_0,B_1)_{\theta,\infty}} =$$

$$= \sup_{t>0} t^{-\theta}K(t,u) < \infty\} \ ; \ q = \infty .$$

$(B_0,B_1)_{\theta,q}$ are Banach spaces.

Complex method [15]. Let (B_0,B_1) be an interpolation couple. In the strip $\{z \mid 0 \leq \mathrm{Re}\, z \leq 1\}$ in the complex plane we consider functions $f(z)$ with

1. $f(z) \in B_0 + B_1$ for $0 \leq \mathrm{Re}\, z \leq 1$, $f(z)$ is uniformly $(B_0 + B_1)$-bounded and $(B_0 + B_1)$-continuous in $0 \leq \mathrm{Re}\, z \leq 1$.

2. $f(z)$ is a $(B_0 + B_1)$-analytic function in $0 < \mathrm{Re}\, z < 1$.

3. (a) $f(it) \in B_0$ for $t \in R_1$, $f(it)$ is uniformly B_0-bounded and B_0-continuous,

(b) $f(1+it) \in B_1$ for $t \in R_1$, $f(1+it)$ is uniformly B_1-bounded and B_1-continuous.

Let be $0 < \theta < 1$. We set

$$[B_0,B_1]_\theta = \{u \mid u \in B_0 + B_1 \text{ , } \exists f(z) \text{ with the above described properties and } f(\theta) = u\} .$$

$[B_0,B_1]_\theta$ is a Banach space with the norm

$$\|u\|_{[B_0,B_1]_\theta} = \inf_{f(\theta)=u} \max\left[\sup_{t\in R_1} \|f(it)\|_{B_0} \text{ , } \sup_{t\in R_1} \|f(1+it)\|_{B_1}\right] .$$

We do not describe properties of the interpolation spaces $[B_0,B_1]_\theta$ and $(B_0,B_1)_{\theta,q}$ and refer to the cited papers, especially to [34, 15, 13], and [35]. We mention only the simple relations

$$(B_0,B_1)_{\theta,q} = (B_1,B_0)_{1-\theta,q} \text{ ; } [B_0,B_1]_\theta = [B_1,B_0]_{1-\theta} \text{ ,}$$

which show that the order of the spaces B_0 and B_1 is not important. Concerning further development of constructive interpolation methods we mention the papers of Peetre [46] (L-method) and Schechter [48] (general complex method). A description of the interpolation theory in Hilbert spaces is given by Lions - Magenes in [33].

1.2. Interpolation theory for spaces of distributions defined in R_n

1.2.1. Definitions. We denote the Euclidean n-space by R_n. $C_0^\infty(R_n)$ is the set of all complex infinitely differentiable functions defined in R_n with compact support in R_n. $S(R_n)$ is the Schwartz space of all complex rapidly decreasing functions in R_n. $S'(R_n)$ is the dual space to $S(R_n)$, the space of tempered distributions. F is the Fourier transformation in $S'(R_n)$, F^{-1} the inverse Fourier transformation.

Let be $-\infty < \sigma < \infty$. For $1 \leq p < \infty$ we set

$$\ell_p^\sigma = \left\{\xi \mid \xi = (\xi_j)_{j=0,1,2,\ldots} \text{ ; } \xi_j \text{ complex; } \|\xi\|_{\ell_p^\sigma} = \left[\sum_{j=0}^{\infty} 2^{j\sigma p} |\xi_j|^p\right]^{\frac{1}{p}} < \infty\right\}$$

and for $p = \infty$

$$\ell_\infty^\sigma = \{\xi \mid \xi = (\xi_j)_{j=0,1,2,\dots} ; \ \xi_j \ \text{complex};$$

$$\|\xi\|_{\ell_\infty^\sigma} = \sup_j 2^{j\sigma} |\xi_j| < \infty\} .$$

Obviously ℓ_p^σ is a Banach space.

We define the spaces $F_{p,q}^s(R_n)$: Let be $-\infty < s < \infty$; $1 < p < < \infty$; $1 < q < \infty$. We set

(1) $$F_{p,q}^s(R_n) = \left\{ f \mid f \in S'(R_n) , \ f \underset{S'}{=} \sum_{j=0}^{\infty} a_j(x) ; \right.$$

$$\|\{a_j\}\|_{L_p(\ell_q^s)} = \left(\int_{R_n} \left[\sum_{j=0}^{\infty} (2^{sj} |a_j(x)|)^q \right]^{\frac{p}{q}} dx \right)^{\frac{1}{p}} < \infty ;$$

$$\operatorname{supp} Fa_j \subset \{\xi \mid 2^{j-1} \leqq |\xi| \leqq 2^{j+1}\} \ \text{for} \ j = 1, 2, \dots ;$$

$$\left. \operatorname{supp} Fa_0 \subset \{\xi \mid |\xi| \leqq 2\} \right\} .$$

$\sum_{j=0}^{\infty} a_j(x) \underset{S'}{=} f$ means that $\sum_{j=0}^{N} a_j(x)$ converges in $S'(R_n)$ to f. supp g denotes the support of the distribution g. We write

(2) $$\|f\|_{F_{p,q}^s} = \inf_{f=\Sigma a_j} \|\{a_j\}\|_{L_p(\ell_q^s)} .$$

We define the spaces $B_{p,q}^s(R_n)$: For $-\infty < s < \infty$; $1 < p < \infty$; $1 \leqq q < \infty$ we set

(3) $$B_{p,q}^s(R_n) = \left\{ f \mid f \in S'(R_n) , \ f \underset{S'}{=} \sum_{j=0}^{\infty} a_j(x) ; \right.$$

$$\|\{a_j\}\|_{\ell_q^s(L_p)} = \left[\sum_{j=0}^{\infty} (2^{sj} \|a_j(x)\|_{L_p})^q \right]^{\frac{1}{q}} < \infty ;$$

$$\operatorname{supp} Fa_j \subset \{\xi \mid 2^{j-1} \leqq |\xi| \leqq 2^{j+1}\} \ \text{for} \ j = 1, 2, \dots ;$$

$$\left. \operatorname{supp} Fa_0 \subset \{\xi \mid |\xi| \leqq 2\} \right\} .$$

($L_p(R_n)$ is the usual space of p-integrable functions with the usual norm.) For $-\infty < s < \infty$; $1 < p < \infty$ and $q = \infty$ we set

$$B_{p,\infty}^s(R_n) = \left\{ f \mid f \in S'(R_n) , \ f \underset{S'}{=} \sum_{j=0}^{\infty} a_j(x) ; \right.$$

$$\|\{a_j\}\|_{\ell^s_\infty(L_p)} = \sup_j 2^{sj} \|a_j(x)\|_{L_p} < \infty ;$$

$$\text{supp } Fa_j \subset \{\xi \mid 2^{j-1} \leq |\xi| \leq 2^{j+1}\} \quad \text{for} \quad j = 1, 2, \ldots ;$$

$$\text{supp } Fa_0 \subset \{\xi \mid |\xi| \leq 2\}\Big\} .$$

We write

$$\|f\|_{B^s_{p,q}} = \inf_{f=\Sigma a_j} \|\{a_j\}\|_{\ell^s_q(L_p)} . \tag{4}$$

Further we set:

(a) For $-\infty < s < \infty$; $1 < p < \infty$;

$$H^s_p(R_n) = F^s_{p,2}(R_n) .$$

(b) For $-\infty < s < \infty$; s = integer; $1 < p < \infty$;

$$W^s_p(R_n) = H^s_p(R_n) \quad (= F^s_{p,2}(R_n)).$$

(c) For $-\infty < s < \infty$; $s \neq$ integer ; $1 < p < \infty$;

$$W^s_p(R_n) = B^s_{p,p}(R_n) .$$

In Section 1.2.3 we describe equivalent norms for the spaces $B^s_{p,q}(R_n)$, $H^s_p(R_n)$, and $W^s_p(R_n)$ for $s \geq 0$. On the basis of these results it is possible to see that the spaces $B^s_{p,q}(R_n)$, $H^s_p(R_n)$ and $W^s_p(R_n)$ have the usual meaning. $W^k_p(R_n)$; $k = 1, 2, \ldots$ are the Sobolev spaces introduced by Sobolev in 1936 [50, 51] . (Sobolev defined the spaces for bounded domains in R_n , but the difference is not important.) $W^0_p(R_n) = L_p(R_n)$. $W^s_p(R_n)$; $0 < s \neq$ integer are the Slobodeckij spaces introduced by Slobodeckij in 1957 [49].
$B^s_{p,q}(R_n)$; $s > 0$ are the Besov spaces introduced by Besov in 1961 [9]. $H^s_p(R_n)$ are the Lebesgue spaces or Bessel potential spaces considered by many authors, see for instance [5]. The definition (3), (4) for the Besov spaces is given in [61]. It is a modification of similar definitions of S. M. Nikol´skij [41] and Peetre [44, 45]. The spaces $F^s_{p,q}(R_n)$ are introduced by the author in [61]. The described definitions for the spaces $B^s_{p,q}(R_n)$ and $F^s_{p,q}(R_n)$ are very convenient for the application of the interpolation theory. All interpolation results for these spaces formulated in Section 1.2.4 are proved on the basis of these definitions in [61].

1.2.2. Inclusion properties, lifting properties, duality, Schauder bases.

Theorem 1. (a) Let be $-\infty < s < \infty$; $1 < p < \infty$; $1 < q < \infty$. $F^s_{p,q}(R_n)$ is a Banach space. $C_0^\infty(R_n)$ is dense in it.

(b) Let be $-\infty < s < \infty$; $1 < p < \infty$; $1 \leq q \leq \infty$. $B^s_{p,q}(R_n)$ is a Banach space. For $q < \infty$, $C_0^\infty(R_n)$ is dense in it.

(c) Let be $-\infty < s < \infty$; $\varepsilon > 0$; $1 < p < \infty$; $1 \leq q_1 \leq q_2 \leq \infty$ and $1 \leq q$, q', $q'' \leq \infty$. Then

(5)
$$B^s_{p,1}(R_n) \subset B^s_{p,q_1}(R_n) \subset B^s_{p,q_2}(R_n) \subset B^s_{p,\infty}(R_n)$$

and

(6)
$$B^{s+\varepsilon}_{p,q'}(R_n) \subset B^s_{p,q}(R_n) \subset B^{s-\varepsilon}_{p,q''}(R_n) .$$

(d) Let be $-\infty < s < \infty$ and $1 < p < \infty$. Then

(7a)
$$B^s_{p,q}(R_n) \subset F^s_{p,q}(R_n) \subset B^s_{p,p}(R_n) \quad \text{for} \quad 1 < q \leq p < \infty$$

and

(7b)
$$B^s_{p,p}(R_n) \subset F^s_{p,q}(R_n) \subset B^s_{p,q}(R_n) \quad \text{for} \quad 1 < p \leq q < \infty .$$

The sign $\subset$ always means that the embedding is continuous. A proof of this theorem is given in [61]. The part (b) is well known, also (c). On the basis of the given definitions it is easy to prove (5) and (6). We remarked in 1.2.1 that $F^s_{p,2}(R_n) = H^s_p(R_n)$ are the usual Lebesgue spaces. In this case (7) means

(8a)
$$B^s_{p,2}(R_n) \subset H^s_p(R_n) \subset B^s_{p,p}(R_n) \quad \text{for} \quad 2 \leq p < \infty$$

and

$$B^s_{p,p}(R_n) \subset H^s_p(R_n) \subset B^s_{p,p}(R_n) \quad \text{for} \quad 1 < p \leq 2 .$$

These relations are also known, see [41],[35] or [52]. The last formulas show

(9)
$$H^s_2(R_n) = B^s_{2,2}(R_n) .$$

Further it follows from the definition of the spaces $F^s_{p,q}(R_n)$ and $B^s_{p,q}(R_n)$ that

(10)
$$F^s_{p,p}(R_n) = B^s_{p,p}(R_n) \; ; \; -\infty < s < \infty \; ; \; 1 < p < \infty$$

holds. This shows that the spaces $F^s_{p,q}(R_n)$ include the spaces

$H^s_p(R_n)$ as well as the spaces $B^s_{p,p}(R_n)$. This is the motive for the introduction of these spaces.

We consider the operation

$$I_s f = F^{-1}(1 + |x|^2)^{\frac{s}{2}} Ff \ ; -\infty < s < \infty .$$

It is easy to see that I_s is a linear continuous one-to-one map from $S(R_n)$ onto $S(R_n)$ and from $S'(R_n)$ onto $S'(R_n)$. It is

$$I_s^{-1} = I_{-s} . \tag{11}$$

Theorem 2. (Lifting property.) Let be $-\infty < s < \infty$; $-\infty < \sigma < \infty$; $1 < p < \infty$.

(a) I_s is a continuous one-to-one operator from $F^{\sigma}_{p,q}(R_n)$ onto $F^{\sigma-s}_{p,q}(R_n)$; $1 < q < \infty$.

(b) I_s is a continuous one-to-one operator from $B^{\sigma}_{p,q}(R_n)$ onto $B^{\sigma-s}_{p,q}(R_n)$; $1 \leqq q \leqq \infty$.

A proof of this theorem is not very hard [61]. For the spaces $B^s_{p,q}(R_n)$ and $H^s_p(R_n) = F^s_{p,2}(R_n)$ the result is known [41]. By means of (11) it follows immediately that the spaces $B^s_{p,q}(R_n)$ for fixed p and q are isomorphic to each other. The same is true for the spaces $F^s_{p,q}(R_n)$ for fixed p and q.

Let B be a Banach space. Then we denote by B' the dual space, the set of all linear continuous functionals with the usual norm. We want to determine $(B^s_{p,q})'$ and $(F^s_{p,q})'$. It is

$$S(R_n) \subset F^s_{p,q}(R_n) \ , \ B^s_{p,q}(R_n) \subset S'(R_n) . \tag{12}$$

Here $S(R_n)$ has the usual local convex topology, and $S'(R_n)$ is the dual space of tempered distributions with the strong topology. $S(R_n)$ is dense in $F^s_{p,q}(R_n)$ as well as in $B^s_{p,q}(R_n)$ with $q < \infty$ (see Theorem 1). So we can interpret in the usual way the dual spaces $(F^s_{p,q}(R_n))'$ and $(B^s_{p,q}(R_n))'$ as subsets of $S'(R_n)$,

$$(F^s_{p,q}(R_n))' \subset S'(R_n) \ ; \quad (B^s_{p,q}(R_n))' \subset S'(R_n) .$$

In this sense we have to understand the next theorem.

Theorem 3. Let be $-\infty < s < \infty$; $1 < p < \infty$; $1 < q < \infty$ and

$$\frac{1}{p} + \frac{1}{p'} = \frac{1}{q} + \frac{1}{q'} = 1 .$$

Then

$$(B^s_{p,q}(R_n))' = B^{-s}_{p',q'}(R_n) \quad \text{and} \quad (F^s_{p,q}(R_n))' = F^{-s}_{p',q'}(R_n) .$$

A proof of this theorem is given in [61]. For the spaces $B^s_{p,q}(R_n)$ the result is known and proved by Taibleson in [53]. It follows that the spaces $B^s_{p,q}(R_n)$ and $F^s_{p,q}(R_n)$ with $1 < p < \infty$ and $1 < q < \infty$ are reflexive. A special case of the last theorem is

$$(H^s_p(R_n))' = H^{-s}_{p'}(R_n) ;$$

$$-\infty < s < \infty ; \; 1 < p < \infty ; \; \frac{1}{p} + \frac{1}{p'} = 1 .$$

Finally we consider the existence of a Schauder basis in $B^s_{p,q}(R_n)$ and $H^s_p(R_n)$. Let B be a complex Banach space. A set of elements $\{u_j\}_{j=1,2,\ldots}$ is called a Schauder basis in B if each element u of B has a unique representation in the form

$$u = \sum_{j=1}^{\infty} \alpha_j u_j , \quad \alpha_j \text{ complex numbers.}$$

<u>Theorem 4</u>. Let be $-\infty < s < \infty$; $1 < p < \infty$; $1 \leq q < \infty$. Then the spaces

$$H^s_p(R_n) \quad \text{and} \quad B^s_{p,q}(R_n)$$

have a Schauder basis.

A proof of this theorem is given in [62] for $s > 0$. But Theorem 2 (lifting property) shows that the same is true for arbitrary real s . In 1.2.1 we remarked $W^0_p(R_n) = H^0_p(R_n) = L_p(R_n)$. For these spaces the result is well known, see [25]. (In [25] spaces $L_p(\Omega)$ are considered where Ω is a set in R_n with bounded measure. But the reasoning in [25] works also for the spaces $L_p(R_n)$.) Theorem 2 shows that $H^s_p(R_n)$ is isomorphic to $L_p(R_n)$. Hence it follows immediately that the spaces $H^s_p(R_n)$ have a Schauder basis. The last theorem shows further that in particular the Sobolev-Slobodeckij spaces $W^s_p(R_n)$ have a Schauder basis.

<u>1.2.3. Equivalent norms</u>. We remarked in 1.2.1 that the spaces $W^s_p(R_n)$, $H^s_p(R_n)$, and $B^s_{p,q}(R_n)$ have the usual sense (at least for $s > 0$). This will be clearer after a new description of these spaces by means of equivalent norms.

Theorem 5. (a) Let be $-\infty < s < \infty$ and $1 < p < \infty$. Then

$$(13) \qquad H_p^s(R_n) = F_{p,2}^s(R_n) = \{f \mid f \in S'(R_n) ;$$

$$g = F^{-1}(1 + |\xi|^2)^{s/2} Ff \in L_p(R_n)\} ,$$

$\|g\|_{L_p}$ is an equivalent norm to $\|f\|_{H_p^s}$.

(b) It is $H_p^0(R_n) = L_p(R_n)$. Let k be an integer; $k = 1, 2, \ldots$. Then

$$(14) \qquad W_p^k(R_n) = H_p^k(R_n) = F_{p,2}^k(R_n) =$$

$$= \{f \mid D^\alpha f \in L_p(R_n) \quad \text{for} \quad |\alpha| \leq k\} .$$

$\sum_{|\alpha|\leq k} \|D^\alpha f\|_{L_p}$ is an equivalent norm to $\|f\|_{W_p^k}$.

(We use the usual notation $D^\alpha = \partial^{|\alpha|}/(\partial x_1^{\alpha_1} \ldots \partial x_n^{\alpha_n})$ for $\alpha = (\alpha_1,\ldots,\alpha_n)$.)

(14) and (13) are the usual definitions for the Sobolev spaces and the Lebesgue spaces. The equivalence to our definition is proved by Nikol´skij [41] and in [61].

We pass on to equivalent norms in the Besov spaces $B_{p,q}^s(R_n)$ with $s > 0$. (These spaces include the Slobodeckij spaces $W_p^s(R_n) = B_{p,p}^s(R_n)$ for $s \neq$ integer.) We write

$$(\Delta_{h,k} f)(x) = f(x_1,\ldots,x_{k-1},x_k + h,x_{k+1},\ldots,x_n) - f(x) ; \quad h \in R_1 ;$$

$$(\Delta_h f)(x) = f(x + h) - f(x) ; \quad h \in R_n$$

and

$$\Delta_{h,k}^\ell = \Delta_{h,k}(\Delta_{h,k}^{\ell-1}) ; \quad \Delta_h^\ell = \Delta_h(\Delta_h^{\ell-1})$$

for $\ell = 2, 3, \ldots$. Further we set for a real number $\varkappa$:

$$\varkappa = [\varkappa] + \{\varkappa\} ; \quad [\varkappa] \text{ integer; } 0 \leq \{\varkappa\} < 1$$

and

$$\varkappa = [\varkappa]^- + \{\varkappa\}^+ ; \quad [\varkappa]^- \text{ integer; } 0 < \{\varkappa\}^+ \leq 1 .$$

Theorem 6. Let be $0 < s < \infty$; $1 < p < \infty$; $1 \leq q \leq \infty$. Then

$$B_{p,q}^s(R_n) = \{f \mid f \in S'(R_n) ; \|f\|_{B_{p,q}^s}^{(r)} < \infty\} ; \quad r = 1, 2, 3$$

where

(a)

$$(15)\qquad \|f\|^{(1)}_{B^s_{p,q}} = \|f\|_{L_p} + \left[\int_0^\infty h^{-(s-j)q} \sum_{k=1}^{n} \left\|\Delta^{\ell}_{h,k}\frac{\partial^j f}{\partial x_k^j}\right\|^q_{L_p} \frac{dh}{h}\right]^{\frac{1}{q}}$$

(with the usual modification for $q = \infty$); j and ℓ are integers; $0 \leq j < s$; $\ell > s - j$.

(b)

$$(16)\qquad \|f\|^{(2)}_{B^s_{p,q}} = \|f\|_{L_p} + \left[\int_0^\infty h^{-(s-j)q} \sum_{k=1}^{n} \sum_{|\alpha|\leq j} \|\Delta^{\ell}_{h,k}D^{\alpha}f\|^q_{L_p} \frac{dh}{h}\right]^{\frac{1}{q}}$$

(with the usual modification for $q = \infty$); j and ℓ are integers; $0 \leq j < s$; $\ell > s - j$.

(c)

$$(17)\qquad \|f\|^{(3)}_{B^s_{p,q}} = \|f\|_{L_p} + \left[\int_{R_n} |h|^{-(s-j)q} \sum_{|\alpha|\leq j} \|\Delta^{\ell}_{h}D^{\alpha}f\|^q_{L_p} \frac{dh}{|h|^n}\right]^{\frac{1}{q}}$$

(with the usual modification for $q = \infty$); j and ℓ are integers; $0 \leq j < s$; $\ell > s - j$.

The norms $\|f\|^{(r)}_{B^s_{p,q}}$; $r = 1, 2, 3$ are equivalent to the norm $\|f\|_{B^s_{p,q}}$.

This theorem shows that $B^s_{p,q}(R_n)$ are the usual Besov spaces, see [9, 41]. A proof of the theorem is given by Nikol´skij [41], see also [61, 64]. The most interesting case of the choice of j and ℓ is $j = [s]^-$, $\ell = 1 + [\{s\}^+]$. By means of this choice of j and ℓ we find for the Slobodeckij spaces $W^s_p(R_n) = B^s_{p,p}(R_n)$; $s \neq$ integer ; $1 < p < \infty$ that

$$(18)\qquad \|f\|^{(3)}_{W^s_p} = \|f\|_{L_p} + \left[\int_{R_n\times R_n} \sum_{|\alpha|\leq[s]} \frac{|D^{\alpha}f(x) - D^{\alpha}f(y)|^p}{|x - y|^{n+\{s\}p}}\, dx\, dy\right]^{\frac{1}{p}} .$$

This is Slobodeckij´s definition [49]. (18) follows immediately from (17). There are many other equivalent norms. So we can replace in (17) $\sum_{|\alpha|\leq j}$ by $\sum_{|\alpha|=j}$ or by a sum only over $D^{\alpha} = \partial^j/\partial x_k^j$ in the same way as in (15). For further equivalent norms we refer to Nikol - skij [41], p. 255 - 256, and Taibleson [52].

In (8) and (9) we described relations between the spaces $H^s_p(R_n)$ and the spaces $B^s_{p,q}(R_n)$. (9) is the only case where $H^s_p(R_n)$ and $B^s_{p,q}(R_n)$ are equal. That means

$$H^s_p(R_n) \neq B^s_{p,q}(R_n) \tag{19}$$

for $-\infty < s < \infty$; $1 < p < \infty$; $1 \leq q \leq \infty$; $|p - 2| + |q - 2| > 0$. For references concerning this problem see [41], p. 438 - 440.

We do not present embedding theorems and extension theorems for the considered spaces. We return to this question in connection with the study of these spaces defined in domains.

1.2.4. Interpolation theory. The main purpose of this paper is the description of interpolation results for spaces of Besov type.

Theorem 7. Let be $-\infty < s_0$, $s_1 < \infty$; $s_0 \neq s_1$; $1 < p < \infty$; $1 \leq q_0$, q_1 , $q \leq \infty$; $0 < \theta < 1$ and $s = (1 - \theta)s_0 + \theta s_1$. Then

$$\begin{aligned}(B^{s_0}_{p,q_0}(R_n),B^{s_1}_{p,q_1}(R_n))_{\theta,q} &= (B^{s_0}_{p,q_0}(R_n),F^{s_1}_{p,q_1}(R_n))_{\theta,q} = \\ &= (F^{s_0}_{p,q_0}(R_n),F^{s_1}_{p,q_1}(R_n))_{\theta,q} = B^s_{p,q}(R_n) .\end{aligned} \tag{20}$$

The formula

$$(B^{s_0}_{p,q_0}(R_n),B^{s_1}_{p,q_1}(R_n))_{\theta,q} = B^s_{p,q}(R_n)$$

is due to Lions-Peetre [34]. Then (20) follows immediately from the inclusion property (7) and the reiteration theorem of the interpolation theory [34]. We mention the particular case of (20)

$$\begin{aligned}(B^{s_0}_{p,q_0}(R_n),H^{s_1}_p(R_n))_{\theta,q} &= (B^{s_0}_{p,q_0}(R_n),W^{s_1}_p(R_n))_{\theta,q} = \\ &= (H^{s_0}_p(R_n),H^{s_1}_p(R_n))_{\theta,q} = (H^{s_0}_p(R_n),W^{s_1}_p(R_n))_{\theta,q} = \\ &= (W^{s_0}_p(R_n),W^{s_1}_p(R_n))_{\theta,q} = B^s_{p,q}(R_n) ,\end{aligned} \tag{21}$$

where the parameters have the same meaning as in Theorem 7.

The spaces in (20) have always the same index p . Roughly spoken the index p determines the "main structure" of the spaces $B^s_{p,q}(R_n)$ and $F^s_{p,q}(R_n)$. Now we pass on to an interpolation theorem for spaces with different "main structures".

Theorem 8. Let be $-\infty < s_0 , s_1 < \infty ; 1 < p_0 , p_1 , q_0 , q_1 < \infty$ and $0 < \theta < 1$. We set

$$\frac{1}{p} = \frac{1-\theta}{p_0} + \frac{\theta}{p_1} ; \quad \frac{1}{q} = \frac{1-\theta}{q_0} + \frac{\theta}{q_1} \quad \text{and} \quad s = (1-\theta)s_0 + \theta s_1 .$$

(a) For $s_0 \neq s_1$, it is

$$(F^{s_0}_{p_0,q_0}(R_n), F^{s_1}_{p_1,q_1}(R_n))_{\theta,p} = B^s_{p,p}(R_n) . \tag{22}$$

(b) For $s_0 = s_1 = s$; $q_0 \neq q_1$ and $p = q$, it is

$$(F^{s}_{p_0,q_0}(R_n), F^{s}_{p_1,q_1}(R_n))_{\theta,p} = B^s_{p,p}(R_n) . \tag{23}$$

(c) For $s_0 = s_1 = s$ and $q_0 = q_1 = q$, it is

$$(F^{s}_{p_0,q}(R_n), F^{s}_{p_1,q}(R_n))_{\theta,p} = F^s_{p,q}(R_n) . \tag{24}$$

(d) It is

$$[B^{s_0}_{p_0,q_0}(R_n), B^{s_1}_{p_1,q_1}(R_n)]_{\theta} = B^s_{p,q}(R_n) \tag{25}$$

and

$$[F^{s_0}_{p_0,q_0}(R_n), F^{s_1}_{p_1,q_1}(R_n)]_{\theta} = F^s_{p,q}(R_n) . \tag{26}$$

This theorem is proved in [61] on the basis of the definition of the spaces $B^s_{p,q}(R_n)$ and $F^s_{p,q}(R_n)$ in 1.2.1. Formula (25) is known and due to Grisvard [21] and Taibleson [53]. We discuss the theorem and consider its special cases.

Using (10) and the parts (a) and (b) of the last theorem we find

$$(B^{s_0}_{p_0,p_0}(R_n), B^{s_1}_{p_1,p_1}(R_n))_{\theta,p} = B^s_{p,p}(R_n) \tag{27}$$

with $-\infty < s_0 , s_1 < \infty ; 1 < p_0 , p_1 < \infty ; 0 < \theta < 1$;

$$\frac{1}{p} = \frac{1-\theta}{p_0} + \frac{\theta}{p_1} \quad \text{and} \quad s = (1-\theta)s_0 + \theta s_1 . \tag{28}$$

This result is due to Grisvard [21], see also Peetre [44]. A discussion of the more general interpolation spaces

$$(B^{s_0}_{p_0,q_0}(R_n), B^{s_1}_{p_1,q_1}(R_n))_{\theta,r}$$

is given by Peetre in [43, 45].

The Lebesgue spaces $H^s_p(R_n)$ are determined by $H^s_p(R_n) = F^s_{p,2}(R_n)$. A special case of (22) is

$$(H^{s_0}_{p_0}(R_n), H^{s_1}_{p_1}(R_n))_{\theta,p} = B^s_{p,p}(R_n) \tag{29}$$

with $-\infty < s_0$, $s_1 < \infty$; $s_0 \neq s_1$; $1 < p_0$, $p_1 < \infty$ and $0 < \theta < 1$; p and s are determined in (28). This result is presented by Peetre [45] without proof.

As a consequence of (24) we obtain

$$(H^s_{p_0}(R_n), H^s_{p_1}(R_n))_{\theta,p} = H^s_p(R_n) \tag{30}$$

with $-\infty < s < \infty$; $1 < p_0$, $p_1 < \infty$, and

$$\frac{1}{p} = \frac{1-\theta}{p_0} + \frac{\theta}{p_1} .$$

The special case $s = 0$ leads to the well-known interpolation result

$$(L_{p_0}(R_n), L_{p_1}(R_n))_{\theta,p} = L_p(R_n) ,$$

[34]. A further consequence of (22) is

$$(H^{s_0}_{p_0}(R_n), B^{s_1}_{p_1,p_1}(R_n))_{\theta,p} = B^s_{p,p}(R_n) \tag{31}$$

with $-\infty < s_0$, $s_1 < \infty$; $s_0 \neq s_1$; $1 < p_0$, $p_1 < \infty$ and $0 < \theta < 1$. p and s are determined in (28).

Now we discuss special cases of Theorem 8 (d). (26) implies

$$[H^{s_0}_{p_0}(R_n), H^{s_1}_{p_1}(R_n)]_\theta = H^s_p(R_n) \tag{32}$$

with $-\infty < s_0$, $s_1 < \infty$; $1 < p_0$, $p_1 < \infty$; $0 < \theta < 1$. s and p are determined in (28). This result is due to Calderón [14]. With the same conditions for the indices we have

$$[H^{s_0}_{p_0}(R_n), B^{s_1}_{p_1,p_1}(R_n)]_\theta = F^s_{p,q}(R_n) , \quad \frac{1}{q} = \frac{1-\theta}{2} + \frac{\theta}{p_1} .$$

1.3. Interpolation theory for function spaces without weights defined in domains

1.3.1. Definitions. We consider two kinds of domains Ω, $\Omega \subset R_n$, with the boundary $\partial\Omega$.

(a) Ω is a bounded domain with a smooth boundary. For the sake of simplicity we assume $\partial\Omega \in C^\infty$. Many results are true by weaker assumptions for the boundary. Some theorems for bounded domains which we shall formulate are true for bounded domains with the restricted cone property in the sense of Agmon [2], p. 11.

(b) Ω is an unbounded domain with the cone property. That means: For a suitable open one-side unbounded cone K it is

$$x + K \subset \Omega \quad \text{for all} \quad x \in \Omega . \tag{33}$$

There is another cone K' with (33) and $\bar{K} \subset \{y \,\big|\, |y| = 1\} \subset K'$.

We define $B^s_{p,q}(\Omega)$ and $H^s_p(\Omega)$ (and so also the special cases $W^s_p(\Omega)$) for $s > 0$ as the restriction of $B^s_{p,q}(R_n)$ and $H^s_p(R_n)$ on Ω. That means:

(a) For $0 < s < \infty$; $1 < p < \infty$, and $1 \leq q \leq \infty$ we define

$$B^s_{p,q}(\Omega) = B^s_{p,q}(R_n) \Big/ \{f \mid f \in B^s_{p,q}(R_n) , \; f(x) = 0 \text{ almost everywhere in } \Omega\} , \tag{34}$$

$$\|f\|_{B^s_{p,q}(\Omega)} = \inf_{\substack{\tilde{f} \in B^s_{p,q}(R_n) \\ \tilde{f}(x) = f(x) \text{ a.e. in } \Omega}} \|\tilde{f}\|_{B^s_{p,q}(R_n)} .$$

(b) For $0 \leq s < \infty$ and $1 < p < \infty$ we define

$$H^s_p(\Omega) = H^s_p(R_n) \Big/ \{f \mid f \in H^s_p(R_n) , \; f(x) = 0 \text{ almost everywhere in } \Omega\} , \tag{35}$$

$$\|f\|_{H^s_p(\Omega)} = \inf_{\substack{\tilde{f} \in H^s_p(R_n) \\ \tilde{f}(x) = f(x) \text{ a.e. in } \Omega}} \|\tilde{f}\|_{H^s_p(R_n)} .$$

Further we set for $s =$ integer and $1 < p < \infty$

$$W^s_p(\Omega) = H^s_p(\Omega) ,$$

and for $s \neq$ integer and $1 < p < \infty$

$$W_p^s(\Omega) = B_{p,p}^s(\Omega) .$$

Obviously $W_p^s(\Omega)$ is the restriction of $W_p^s(R_n)$ on Ω. $W_p^s(\Omega)$ are the Sobolev-Slobodeckij spaces, $H_p^s(\Omega)$ the Lebesgue spaces, and $B_{p,q}^s(\Omega)$ the Besov spaces for domains. For historical remarks about these spaces see 1.2.1. We could also define the spaces $F_{p,q}^s(\Omega)$, but we do not do it. For our purpose the special case $H_p^s(\Omega) = F_{p,2}^s(\Omega)$ will be sufficient. In section 1.3.2 we give an "inner" description of the spaces $W_p^s(\Omega)$ and $B_{p,q}^s(\Omega)$ using only the properties of the functions in Ω. If the boundary of the domain is not smooth it is necessary to start with the "inner" description of the spaces $W_p^s(\Omega)$ and $B_{p,q}^s(\Omega)$. The Besov spaces for such general domains are considered by V. P. Il'in [23] and Muramatu [38]. Here we always assume the smoothness properties for the domains described above.

Let $C_0^\infty(\Omega)$ be the set of all complex infinitely differentiable functions with compact support in Ω. Then we denote by $\overset{\circ}{B}{}_{p,q}^s(\Omega)$ $(\overset{\circ}{H}{}_p^s(\Omega))$ the completion of $C_0^\infty(\Omega)$ in $B_{p,q}^s(\Omega)$ $(H_p^s(\Omega))$.

1.3.2. Equivalent norms. It is possible to describe the Sobolev-Slobodeckij spaces $W_p^s(\Omega)$ and the Besov spaces $B_{p,q}^s(\Omega)$ in a more direct way using only the properties of the functions in Ω. For the Lebesgue spaces $H_p^s(\Omega)$, $s \neq$ integer, this seems to be somewhat complicated.

If Ω is a domain in R_n and $h \in R_n$ we shall write

$$\Omega_{h,\ell} = \bigcap_{j=0}^{\ell} \{x \mid x + jh \in \Omega\} .$$

Theorem 9. Let Ω be an unbounded domain with the cone property and let K be a cone with (33). Let be $0 < s < \infty$; $1 < p < \infty$; $1 \leq q \leq \infty$.

(a) It is $H_p^0(\Omega) = L_p(\Omega)$. For $s =$ integer, it is

$$(36)\quad W_p^s(\Omega) = H_p^s(\Omega) = \left\{f \mid f \in L_p(\Omega),\ \|f\|_{W_p^s}^* = \sum_{|\alpha|\leq s} \|D^\alpha f\|_{L_p} < \infty\right\},$$

$\|f\|_{W_p^s}^*$ is an equivalent norm to $\|f\|_{W_p^s}$.

(b) It is

$$B_{p,q}^s(\Omega) = \{f \mid f \in L_p(\Omega),\ \|f\|_{B_{p,q}^s}^{(r)} < \infty\};\quad r = 1, 2, 3$$

where

$$(37)\quad \|f\|^{(1)}_{B^s_{p,q}} = \|f\|_{L_p(\Omega)} + \left[\int_K |h|^{-(s-j)q} \sum_{k=1}^{n} \left\|\Delta_h^\ell \frac{\partial^j f}{\partial x_k^j}\right\|^q_{L_p(\Omega)} \frac{dh}{|h|^n}\right]^{\frac{1}{q}},$$

$$(38)\quad \|f\|^{(2)}_{B^s_{p,q}} = \|f\|_{L_p(\Omega)} + \left[\int_K |h|^{-(s-j)q} \sum_{|\alpha|\leq j} \|\Delta_h^\ell D^\alpha f\|^q_{L_p(\Omega)} \frac{dh}{|h|^n}\right]^{\frac{1}{q}},$$

$$(39)\quad \|f\|^{(3)}_{B^s_{p,q}} = \|f\|_{L_p(\Omega)} + \left[\int_{R_n} |h|^{-(s-j)q} \sum_{|\alpha|\leq j} \|\Delta_h^\ell D^\alpha f\|^q_{L_p(\Omega_{h,\ell})} \frac{dh}{|h|^n}\right]^{\frac{1}{q}}$$

with the usual modification for $q = \infty$; j and ℓ are integers; $0 \leq j < s$; $\ell > s - j$. The norms $\|f\|^{(r)}_{B^s_{p,q}}$ are equivalent to the norm $\|f\|_{B^s_{p,q}(\Omega)}$.

A proof of this theorem is given in [64]. The theorem is the extension of Theorem 6 on domains in R_n . (36) is the usual definition of the Sobolev spaces. Formula (39) coincides with Muramatu's definition [38] (we can replace $\sum_{|\alpha|\leq j}$ by $\sum_{|\alpha|=j}$ or by the sum over the terms $D^\alpha f = \partial^j f/\partial x_k^j$; $k = 1, \ldots, n$). See also Il'in [23]. The most interesting case of the choice of j and ℓ is again $j = [s]^-$ and $\ell = 1 + [\{s\}^+]$. Particularly, we have for the Slobodeckij spaces $W^s_p(\Omega) = B^s_{p,p}(\Omega)$; $s \neq$ integer ; $1 < p < \infty$;

$$(40)\quad \|f\|^{(3)}_{W^s_p} = \|f\|_{L_p} + \left[\int_{\Omega\times\Omega} \sum_{|\alpha|\leq[s]} \frac{|D^\alpha f(x) - D^\alpha f(y)|^p}{|x-y|^{n+\{s\}p}} dx\, dy\right]^{\frac{1}{p}} .$$

Now we pass on to bounded domains Ω with $\partial\Omega \in C^\infty$. For $\lambda > 0$ we set

$$\Omega^{(\lambda)} = \{x \mid x \in \Omega,\ d(x) > \lambda\} ,$$

$d(x)$ being the distance to the boundary,

$$d(x) = \inf_{y\in\partial\Omega} |x - y| .$$

For $0 < \lambda \leq \lambda_0$ it is

$$\Omega - \Omega^{(\lambda)} = \partial\Omega \times (0,\lambda] ,$$

where $(0,\lambda]$ is taken in the direction of the inner normal ν_z, $z \in \partial\Omega$, $|\nu_z| = 1$. For $h \in R_n$, $h \neq 0$; $0 < \varepsilon < \pi/2$ and $0 < \lambda \leq \lambda_0$ we write

$$\Omega_h = \Omega_{h,\varepsilon,\lambda} = \Omega^{(\lambda)} \cup (\{z \mid z \in \partial\Omega,\ 0 \leq \langle h,\nu_z\rangle < \varepsilon\} \times (0,\lambda]) , \tag{41}$$

where $\langle h,\nu_z\rangle$ denotes the angle between the directions h and ν_z. This means that Ω_h is the union of $\Omega^{(\lambda)}$ and that part of $\Omega - \Omega^{(\lambda)}$ for which the directions ν_z and h are "near" each other. The construction (41) is helpful for the interpolation theory of function spaces with weights. But to give the possibility to compare we shall use it also here.

Theorem 10. Let Ω be a bounded domain with $\partial\Omega \in C^\infty$. Let be $0 < s < \infty$; $1 < p < \infty$; $1 \leq q \leq \infty$.

(a) It is $H^0_p(\Omega) = L_p(\Omega)$. For s = integer (36) holds, $\|f\|^*_{W^s_p}$ is equivalent to the norm $\|f\|_{W^s_p}$.

(b) It is

$$B^s_{p,q}(\Omega) = \{f \mid f \in L_p(\Omega) , \|f\|^{(r)}_{B^s_{p,q}} < \infty\} ; \quad r = 1,\ 2,\ 3$$

where

$$\|f\|^{(1)}_{B^s_{p,q}} = \|f\|_{L_p(\Omega)} + \left[\int_{|h|\leq\delta} |h|^{-(s-j)q} \sum_{k=1}^{n} \left\|\Delta_h^\ell \frac{\partial^j f}{\partial x_k^j}\right\|^q_{L_p(\Omega_h)} \frac{dh}{|h|^n}\right]^{\frac{1}{q}} \tag{42}$$

($\delta > 0$ sufficiently small; ε and λ in (41) are fixed numbers),

$$\|f\|^{(2)}_{B^s_{p,q}} = \|f\|_{L_p(\Omega)} + \left[\int_{|h|\leq\delta} |h|^{-(s-j)q} \sum_{|\alpha|\leq j} \|\Delta_h^\ell D^\alpha f\|^q_{L_p(\Omega_h)} \frac{dh}{|h|^n}\right]^{\frac{1}{q}} . \tag{43}$$

$\|f\|^{(3)}_{B^s_{p,q}}$ has the same meaning as in (39). (For $q = \infty$ we have to modify in the usual way.) j and ℓ are integers; $0 \leq j < s$; $\ell > s - j$.

We can do the same remarks as after Theorem 9; in particular (40) is true.

1.3.3. Embedding theorems and extension theorems, the spaces $\overset{\circ}{B}{}^s_{p,q}$ and $\overset{\circ}{H}{}^s_p$. In this section we give a brief and sketchy survey of the main embedding theorems and extension theorems. We assume that Ω is a bounded domain with $\partial\Omega \in C^\infty$. We denote by $C^\infty(\bar{\Omega})$ the set of all complex functions infinitely differentiable in Ω whose derivatives can be extended continuously on $\bar{\Omega}$ (closure of Ω). For $y \in \partial\Omega$ we denote by $\nu = \nu_y$ the inner normal.

First we define the Banach space $B^s_{p,q}(\partial\Omega)$; $s > 0$; $1 < p < \infty$; $1 \leq q \leq \infty$. We choose $y^{(i)} \in \partial\Omega$, $O_i \subset \partial\Omega$, and $O_i' \subset \partial\Omega$; $i = 1, \ldots, N$ so that

(a) O_i and O_i' are open subsets of $\partial\Omega$, $\bar{O}_i \subset O_i'$, $\bigcup_{i=1}^{N} O_i = \partial\Omega$.

(b) $y^{(i)} \in O_i$. It exists a local system of coordinates $\{z_1,\ldots,z_n\}$ with the origin $y^{(i)}$, the tangential directions $z_1,\ldots,z_{n-1}$ and the normal direction z_n, so that $O_i' \cap \partial\Omega$ has the representation

$$z_n = \varphi^{(i)}(z_1,\ldots,z_{n-1}) , \quad \varphi^{(i)} \text{ is a } C^\infty\text{-function}, \tag{44}$$

$$\varphi^{(i)}(0) = 0 , \quad \frac{\partial\varphi^{(i)}}{\partial z_k}(0) = 0 ; \quad k = 1, \ldots, n-1 .$$

Let P_i be the projection of O_i' on the $\{z_1,\ldots,z_{n-1}\}$-plane. We choose a set of functions $\psi_i \in C^\infty(\partial\Omega)$; $i = 1, \ldots, N$ with

$$0 \leq \psi_i \leq 1 ; \quad \operatorname{supp} \psi_i \subset O_i' , \quad \psi_i(x) = 1 \text{ for } x \in O_i , \tag{45}$$

$$\sum_{i=1}^{N} \psi_i(x) = 1 \text{ for } x \in \partial\Omega .$$

Then we define

$$B^s_{p,q}(\partial\Omega) = \{f \mid f \in L_p(\partial\Omega) ,$$

$$(\psi_i f)(z_1,\ldots,z_{n-1}) \in B^s_{p,q}(P_i) \text{ for } i = 1, \ldots, N\} ,$$

$$\|f\|_{B^s_{p,q}(\partial\Omega)} = \sum_{i=1}^{N} \|\psi_i f\|_{B^s_{p,q}(P_i)} , \tag{46}$$

$0 < s < \infty$; $1 < p < \infty$; $1 \leq q \leq \infty$. It is not difficult to prove that all systems of functions ψ_i (and systems of sets O_i and O_i') lead to the same set $B^s_{p,q}(\partial\Omega)$, and all norms (46) are equivalent to each other.

<u>Theorem 11</u>. Let Ω be a bounded domain with $\partial\Omega \in C^\infty$. Let be $1 < p < \infty$; $1 \leq q \leq \infty$.

(a) For $0 \leq s < \infty$, $C^\infty(\bar{\Omega})$ is dense in $H^s_p(\Omega)$. For $s > \frac{1}{p}$,

$$Kf = \left\{ f, \frac{\partial f}{\partial \nu}, \ldots, \frac{\partial^{[s-1/p]^-}}{\partial \nu^{[s-1/p]^-}} f \right\} \tag{47}$$

is a linear and continuous map from $H^s_p(\Omega)$ onto

$$\prod_{k=0}^{[s-1/p]^-} B^{s-1/p-k}_{p,p}(\partial\Omega) . \tag{48}$$

(b) For $0 < s < \infty$, $C^\infty(\bar{\Omega})$ is dense in $B^s_{p,q}(\Omega)$. For $s > \frac{1}{p}$,

$$Kf = \left\{ f, \frac{\partial f}{\partial \nu}, \ldots, \frac{\partial^{[s-1/p]^-}}{\partial \nu^{[s-1/p]^-}} f \right\}$$

is a linear and continuous map from $B^s_{p,q}(\Omega)$ onto

$$\prod_{k=0}^{[s-1/p]^-} B^{s-1/p-k}_{p,q}(\partial\Omega) . \tag{49}$$

This is a very important embedding theorem. It is the basis for boundary value problems for partial differential equations. A proof was given by Nikol´skij [41], pp. 291, 293, 420, 425. (He considered the case $B^s_{p,q}(R_n)$ and $H^s_p(R_n)$ and the embedding on $\{x \mid x_n = 0\}$, but the difference from the case considered here is not important.) However, the result is due to Besov [9], Nikol´skij, Uspenskij, Stein, Lions-Magenes [29], Magenes [35] and other mathematicians. For references see [41] or [35]. Let us explain the theorem. For $f \in C^\infty(\bar{\Omega})$, (48) and (49) mean

$$\sum_{k=0}^{[s-1/p]^-} \left\| \frac{\partial^k f}{\partial \nu^k} \right\|_{B^{s-1/p-k}_{p,p}(\partial\Omega)} \leq c \|f\|_{H^s_p(\Omega)} \tag{50}$$

and

$$\sum_{k=0}^{[s-1/p]^-} \left\| \frac{\partial^k f}{\partial \nu^k} \right\|_{B^{s-1/p-k}_{p,q}(\partial\Omega)} \leq c \|f\|_{B^s_{p,q}(\Omega)} . \tag{51}$$

For arbitrary $f \in H^s_p(\Omega)$ (or $f \in B^s_{p,q}(\Omega)$) the boundary values $\partial^k f/\partial\nu^k|_{\partial\Omega}$ are determined by means of the previous formulas by continuity in the Banach space $B^{s-1/p-k}_{p,p}(\partial\Omega)$ (or $B^{s-1/p-k}_{p,q}(\partial\Omega)$). [We remarked in the theorem that $C^\infty(\bar{\Omega})$ is dense.] The result is also an "extension theorem" because the map is "onto". It is possible to show that there exists a closed subspace of $H_p^s(\Omega)$ $(B_{p,q}^s(\Omega))$ on which K is a one-to-one map from this subspace onto

$$\prod_{k=0}^{[s-1/p]^-} B_{p,p}^{s-1/p-k}(\partial\Omega) \quad \left(\prod_{k=0}^{[s-1/p]^-} B_{p,q}^{s-1/p-k}(\partial\Omega) \right) .$$

We remark that the last theorem is also a theorem for the Sobolev-Slobodeckij spaces $W_p^s(\Omega)$ because these spaces are special cases of the spaces $H_p^s(\Omega)$ and $B_{p,p}^s(\Omega)$.

Now we discuss the question if Theorem 11 gives a complete description of the boundary values of functions belonging to $H_p^s(\Omega)$ or $B_{p,q}^s(\Omega)$. If $f \in H_p^s(\Omega)$ $(B_{p,q}^s(\Omega))$ and $g \in \overset{0}{H}{}_p^s(\Omega)$ $(\overset{0}{B}{}_{p,q}^s(\Omega))$ then

$$Kf = Kg , \tag{52}$$

where K is the operator defined in the last theorem. We ask if the opposite statement is true: f and g belong to $H_p^s(\Omega)$ $(B_{p,q}^s(\Omega))$ and (52) holds. Is $f - g \in \overset{0}{H}{}_p^s(\Omega)$ $(\overset{0}{B}{}_{p,q}^s(\Omega))$? The following theorem gives the affirmative answer for $H_p^s(\Omega)$ and $B_{p,q}^s(\Omega)$ with $q > 1$ and the negative answer for $B_{p,1}^s(\Omega)$.

Theorem 12. Let Ω be a bounded domain with $\partial\Omega \in C^\infty$. Let be $0 < s < \infty$; $1 < p < \infty$.

(a) For $s > \frac{1}{p}$ and $1 < q \leq \infty$ it is

$$\overset{0}{H}{}_p^s(\Omega) = \left\{ f \mid f \in H_p^s(\Omega) , \left. \frac{\partial^k f}{\partial\nu^k} \right|_{\partial\Omega} = 0 \right.$$
$$\left. \text{for } k = 0, \ldots, [s - \tfrac{1}{p}]^- \right\},$$

$$\overset{0}{B}{}_{p,q}^s(\Omega) = \left\{ f \mid f \in B_{p,q}^s(\Omega) , \left. \frac{\partial^k f}{\partial\nu^k} \right|_{\partial\Omega} = 0 \right.$$
$$\left. \text{for } k = 0, \ldots, [s - \tfrac{1}{p}]^- \right\}.$$

For $0 < s \leq \frac{1}{p}$ and $1 < q \leq \infty$ it is

$$\overset{0}{H}{}_p^s(\Omega) = H_p^s(\Omega) , \quad \overset{0}{B}{}_{p,q}^s(\Omega) = B_{p,q}^s(\Omega) .$$

(b) For $s \geq \frac{1}{p}$ it is

$$\left\| \frac{\partial^{[s-1/p]} f}{\partial\nu^{[s-1/p]}} \right\|_{L_p(\partial\Omega)} \leq c \|f\|_{B_{p,1}^s(\Omega)}$$

and

$$\overset{o}{B}{}^{s}_{p,1}(\Omega) = \left\{ f \mid f \in B^{s}_{p,1}(\Omega) \ , \ \frac{\partial^k f}{\partial \nu^k}\bigg|_{\partial\Omega} = 0 \right.$$

$$\left. \text{for } k = 0, \ldots, [s - \tfrac{1}{p}] \right\} .$$

For $0 < s < \frac{1}{p}$ it is

$$\overset{o}{B}{}^{s}_{p,1}(\Omega) = B^{s}_{p,1}(\Omega) .$$

This theorem is proved in [63]. Special cases are known. For $p = 2$ and the spaces $H^s_2(\Omega) = W^s_2(\Omega)$ the result is due to Lions-Magenes [33]. For $p \neq 2$ and for the spaces $W^s_p(\Omega)$, $H^s_p(\Omega)$, and $B^s_{p,p}(\Omega)$ the result is proved for the "non-singular" values $s - \frac{1}{p} \neq$ $\neq$ integer , see [31] and [35]. The last theorem shows that the behaviour of the spaces $B^s_{p,1}(\Omega)$ is different from the behaviour of the spaces $B^s_{p,q}(\Omega)$ with $q > 1$. This fact is known. For instance let be $s - \frac{n}{p} = \ell$ an integer. Then Besov proved in [10]

$$B^s_{p,1}(\Omega) \subset C^{\ell}(\bar{\Omega})$$

but the same relation for $B^s_{p,q}(\Omega)$ with $q > 1$ is not true [10].

Finally, we formulate another embedding theorem.

Theorem 13. Let Ω be a bounded domain with $\partial\Omega \in C^\infty$. Let be $1 < p_1 , p_2 < \infty$ and $1 \leqq q \leqq \infty$.

(a) For $0 \leqq s_2 \leqq s_1 < \infty$ and $s_1 - \frac{n}{p_1} \geqq s_2 - \frac{n}{p_2}$ it is

$$H^{s_2}_{p_2}(\Omega) \supset H^{s_1}_{p_1}(\Omega) . \tag{53}$$

For $s_1 - \frac{n}{p_1} > s_2 - \frac{n}{p_2}$, the embedding operator is compact. ($H^0_p = L_p$.)

(b) For $0 < s_2 \leqq s_1 < \infty$ and $s_1 - \frac{n}{p_1} \geqq s_2 - \frac{n}{p_2}$ it is

$$B^{s_2}_{p_2,q}(\Omega) \supset B^{s_1}_{p_1,q}(\Omega) . \tag{54}$$

For $s_1 - \frac{n}{p_1} > s_2 - \frac{n}{p_2}$ the embedding operator is compact.

A proof of this theorem was given by Nikol'skij [41], pp. 279, 435. The result is due to Sobolev, Il'in, Besov, Nikol'skij, Stein and other mathematicians. A special case of (53) is Sobolev's well-known embedding theorem for the spaces $W^s_p(\Omega) = H^s_p(\Omega)$; s = integer [50, 51]. An investigation of embedding theorems on the basis of abstract interpolation theory is given by Yoshikawa [66]. See also Peetre [43].

1.3.4. Inclusion properties, isomorphic properties, Schauder bases. In this section Ω is a bounded domain with $\partial\Omega \in C^\infty$. We describe some results similar to the theorems in Section 1.2.2.

Theorem 14. Let Ω be a bounded domain with $\partial\Omega \in C^\infty$. After replacing R_n by Ω the inclusion properties (5), (6), and (8) are true for $s > 0$ and $s - \varepsilon > 0$.

The proof of this theorem is simple. It follows from the definition of the spaces $H_p^s(\Omega)$ and $B_{p,q}^s(\Omega)$ and the formulas (5), (6), and (8). The theorem is also true for unbounded domains with the cone property.

Some authors consider spaces $W_p^s(\Omega)$ with $s < 0$ defined by

$$W_p^s(\Omega) = (\overset{\circ}{W}{}_{p'}^{-s}(\Omega))' , \quad \frac{1}{p} + \frac{1}{p'} = 1 .$$

However, we do not go into details here and refer to Berezanskij [8] and Lions-Magenes [33] for the case of Hilbert spaces and to Lions-Magenes [29, 30, 31, 32] and Magenes [35] for the general case.

Now we formulate a theorem about the existence of Schauder bases. A closed subset of a Banach space is called complemented if it is the range of a continuous projection operator.

Theorem 15. Let Ω be a bounded domain with $\partial\Omega \in C^\infty$.

(a) Let be $0 < s < \infty$; $0 \leq \sigma < \infty$; $1 < p < \infty$; $1 \leq q < \infty$. The spaces $B_{p,p}^s(\Omega)$, $\overset{\circ}{B}{}_{p,p}^s(\Omega)$, $H_p^\sigma(\Omega)$, and $\overset{\circ}{H}{}_p^\sigma(\Omega)$ (and hence also the spaces $W_p^\sigma(\Omega)$ and $\overset{\circ}{W}{}_p^\sigma(\Omega)$) have a Schauder basis. They are isomorphic to a complemented subspace of $L_p((0,1))$.

(b) For fixed p and q with $1 < p < \infty$ and $1 \leq q \leq \infty$ the spaces $B_{p,q}^s(\Omega)$ are isomorphic to each other.

The theorem is proved in [62]. There are other results in this direction. For the Sobolev spaces $W_p^s(\Omega)$ and $\overset{\circ}{W}{}_p^s(\Omega)$; s = integer the result is due to Fučík, John and Nečas [16]. Whether the Besov spaces $B_{p,q}^s(\Omega)$ with $q < \infty$ have a Schauder basis seems to be an open question.

1.3.5. Interpolation theory. The interpolation theory for the Sobolev spaces, Slobodeckij spaces, Lebesgue spaces and Besov spaces defined in domains has been developed by Lion-Magenes [29, 30, 31, 32, 33, 35], Grisvard [22], Muramatu [38], Fujiwara [17] and other mathematicians.

We start with a theorem which is a partial generalization of Theorems 7 and 8. The Sobolev-Slobodeckij spaces $W_p^s(\Omega)$ are special cases of the Lebesgue spaces $H_p^s(\Omega) = W_p^s(\Omega)$ for s = integer and the Besov spaces $B_{p,p}^s(\Omega) = W_p^s(\Omega)$ for $s \neq$ integer. So we formulate the interpolation theorem only for the spaces $H_p^s(\Omega)$ and $B_{p,q}^s(\Omega)$.

Theorem 16. Let Ω be an unbounded domain with the cone property or a bounded domain with $\partial\Omega \in C^\infty$.

(a) Let be $0 < s_0, s_1 < \infty$; $s_0 \neq s_1$; $1 < p < \infty$; $1 \leq q_0, q_1, q \leq \infty$ and $0 < \theta < 1$. Then

$$(B_{p,q_0}^{s_0}(\Omega), B_{p,q_1}^{s_1}(\Omega))_{\theta,q} = B_{p,q}^{s_0(1-\theta)+s_1\theta}(\Omega). \tag{55}$$

(b) Let be $0 < s_0 < \infty$; $0 \leq s_1 < \infty$; $s_0 \neq s_1$; $1 < p < \infty$; $1 \leq q_0, q \leq \infty$ and $0 < \theta < 1$. Then

$$(B_{p,q_0}^{s_0}(\Omega), H_p^{s_1}(\Omega))_{\theta,q} = B_{p,q}^{s_0(1-\theta)+s_1\theta}(\Omega). \tag{56}$$

(c) Let be $0 \leq s_0, s_1 < \infty$; $s_0 \neq s_1$; $1 < p < \infty$; $1 \leq q \leq \infty$ and $0 < \theta < 1$. Then

$$(H_p^{s_0}(\Omega), H_p^{s_1}(\Omega))_{\theta,q} = B_{p,q}^{s_0(1-\theta)+s_1\theta}(\Omega). \tag{57}$$

(d) Let be $0 < s_0, s_1 < \infty$; $1 < p_0, p_1 < \infty$; $0 < \theta < 1$; s and p are determined in (28). Then

$$(B_{p_0,p_0}^{s_0}(\Omega), B_{p_1,p_1}^{s_1}(\Omega))_{\theta,p} = B_{p,p}^{s}(\Omega). \tag{58}$$

(e) Let be $0 < s_0 < \infty$; $0 \leq s_1 < \infty$; $s_0 \neq s_1$; $1 < p_0, p_1 < \infty$; $0 < \theta < 1$; p and s are determined in (28). Then

$$(B_{p_0,p_0}^{s_0}(\Omega), H_{p_1}^{s_1}(\Omega))_{\theta,p} = B_{p,p}^{s}(\Omega). \tag{59}$$

(f) Let be $0 \leq s_0, s_1 < \infty$; $s_0 \neq s_1$; $1 < p_0, p_1 < \infty$; $0 < \theta < 1$; p and s are determined in (28). Then

$$(H_{p_0}^{s_0}(\Omega), H_{p_1}^{s_1}(\Omega))_{\theta,p} = B_{p,p}^{s}(\Omega). \tag{60}$$

(g) Let be $0 \leq s < \infty$; $1 < p_0, p_1 < \infty$; $0 < \theta < 1$; p is determined in (28). Then

$$(H_{p_0}^{s}(\Omega), H_{p_1}^{s}(\Omega))_{\theta,p} = H_p^{s}(\Omega). \tag{61}$$

(h) Let be $0 < s_0, s_1 < \infty$; $1 < p_0, p_1, q_0, q_1 < \infty$; $0 < \theta < 1$, p and s are determined in (28),

$$\frac{1}{q} = \frac{1-\theta}{q_0} + \frac{\theta}{q_1}.$$

Then

$$[B^{s_0}_{p_0,q_0}(\Omega), B^{s_1}_{p_1,q_1}(\Omega)]_\theta = B^s_{p,q}(\Omega) . \tag{62}$$

(i) Let be $0 \leq s_0, s_1 < \infty$; $1 < p_0, p_1 < \infty$; $0 < \theta < 1$, p and s are determined in (28). Then

$$[H^{s_0}_{p_0}(\Omega), H^{s_1}_{p_1}(\Omega)]_\theta = H^s_p(\Omega) . \tag{63}$$

This theorem is a generalization of Theorem 7 and formulas (21), (25), (27), (29), (30), (31) and (32). A proof is given in [64]. The proof is based on the following scheme. The considered domains have the extension property. That means that there exists a linear and continuous operator S acting from $B^s_{p,q}(\Omega)$ $(H^s_p(\Omega))$ into $B^s_{p,q}(R_n)$ $(H^s_p(R_n))$; $s \leq s_0$. This is a consequence of Calderón's extension method, see [2], p. 171. Further we have the restriction operator R,

$$(Rf)(x) = f(x) \text{ almost everywhere in } \Omega ,$$

which is linear and continuous from $B^s_{p,q}(R_n)$ $(H^s_p(R_n))$ onto $B^s_{p,q}(\Omega)$ $(H^s_p(\Omega))$. Then the theorem follows from 1.2.4 and the general interpolation theory. The main difficulty is an "inner" description of the spaces $B^s_{p,q}(\Omega)$ and $W^s_p(\Omega)$. But this we did in 1.2.3. We remark that SR is a projection operator in $B^s_{p,q}(R_n)$ and $H^s_p(R_n)$. This implies that $B^s_{p,q}(\Omega)$ $(H^s_p(\Omega))$ is isomorphic to a complemented subspace of $B^s_{p,q}(R_n)$ $(H^s_p(R_n))$, see [62].

Now we formulate some interpolation results for the spaces $\overset{o}{H}{}^s_p(\Omega)$ and $\overset{o}{B}{}^s_{p,q}(\Omega)$ (and so also for $\overset{o}{W}{}^s_p(\Omega)$). Let B_1 and B_2 be two such spaces and $\overset{o}{B}_1$ and $\overset{o}{B}_2$ the completion of $C^\infty_0(\Omega)$ in these spaces. Then the general interpolation theory implies

$$(\overset{o}{B}_1, \overset{o}{B}_2)_{\theta,q} \subset (B_1, B_2)^{o}_{\theta,q} \tag{64}$$

where $(B_1, B_2)^{o}_{\theta,q}$ denotes the completion of $C^\infty_0(\Omega)$ in $(B_1, B_2)_{\theta,q}$. The assertion that the both spaces in (64) are equal is not true in general. The situation is more complicated.

For a domain Ω we use the notation $\Omega_{h,\ell}$ and Ω_h described in the beginning of 1.3.2 and in (41). Further, we write for a number $\lambda > 0$

$$\Omega^{\lambda} = \{x \mid x \in \Omega \ , \ d(x) < \lambda\} \tag{65}$$

(in our previous notation $\Omega^{\lambda} = \Omega - \Omega^{(\lambda)}$, see (41)).

Theorem 17. Let Ω be an unbounded domain with the cone property or a bounded domain with $\partial\Omega \in C^{\infty}$. Let m_1 and m_2 be two integers; $0 \leqq m_1 < m_2 < \infty$; $1 < p < \infty$; $1 \leqq q \leqq \infty$; $0 < \theta < 1$; $s = (1 - \theta)m_1 + \theta m_2$. Then

$$(\overset{\circ}{W}{}_p^{m_1}(\Omega), \overset{\circ}{W}{}_p^{m_2}(\Omega))_{\theta,q} = \Big\{ f \mid f \in L_p(\Omega) \ ; \tag{66}$$

$$\|f\|^* = \|f\|^{(3)}_{B^s_{p,q}} + \Big[\int_0^{\infty} (t^{-s}\|f\|_{L_p(\Omega^t)})^q \frac{dt}{t}\Big]^{\frac{1}{q}} < \infty \Big\}$$

(with the usual modification for $q = \infty$). $\|f\|^{(3)}_{B^s_{p,q}}$ is described in (39); $0 \leqq j < s$; $\ell > s - j$. If Ω is a bounded domain with $\partial\Omega \in C^{\infty}$ then we can replace $\|f\|^{(3)}_{B^s_{p,q}}$ by $\|f\|^{(2)}_{B^s_{p,q}}$ in the last formula, $\|f\|^{(2)}_{B^s_{p,q}}$ being determined in (43). $\|f\|^*$ is an equivalent norm in $(\overset{\circ}{W}{}_p^{m_1}(\Omega), \overset{\circ}{W}{}_p^{m_2}(\Omega))_{\theta,q}$.

A proof of this theorem is given in [64]. We can also replace $\|f\|^{(3)}_{B^s_{p,q}}$ in (66) by the equivalent norms determined in (37), (38) or (42). Further, it is easy to see that we can replace $\int_0^{\infty} \dots dt$ in (66) by $\int_0^{\sigma} \dots dt$ where σ is an arbitrary positive number. If $p = q$, we obtain

$$\int_0^{\infty} t^{-sp} \|f\|^p_{L_p(\Omega^t)} \frac{dt}{t} = \int_{\Omega} d^{-sp}(x) \, |f(x)|^p \, dx \ . \tag{67}$$

In this interesting case it follows

$$(\overset{\circ}{W}{}_p^{m_1}(\Omega), \overset{\circ}{W}{}_p^{m_2}(\Omega))_{\theta,p} = \Big\{ f \mid f \in L_p(\Omega) \ ; \tag{68}$$

$$\|f\|^{(3)}_{B^s_{p,p}} + \Big[\int_{\Omega} d^{-sp}(x) \, |f(x)|^p \, dx\Big]^{\frac{1}{p}} < \infty \Big\} .$$

This interpolation result is known for bounded domains with $\partial\Omega \in C^{\infty}$. For $p = 2$ (Hilbert space case) see Lions-Magenes [33], for the general case see Lions-Magenes [30] and Grisvard [22]. We return to the question if the two spaces in (64) are equal. (66) shows that

$$(\overset{o}{W}{}_p^{m_1}(\Omega), \overset{o}{W}{}_p^{m_2}(\Omega))_{\theta,q} = \overset{o}{B}{}^s_{p,q}(\Omega) \tag{69}$$

iff

$$\left[\int_0^\infty (t^{-s}\|f\|_{L_p(\Omega^t)})^q \frac{dt}{t}\right]^{\frac{1}{q}} \le c\|f\|_{B^s_{p,q}} \quad \text{for} \quad f \in \overset{o}{B}{}^s_{p,q}(\Omega) . \tag{70}$$

Theorem 12 (a) implies that for the "singular" values s = integer + $+ \frac{1}{p}$ and $1 < q < \infty$ (70) is not true. It is not known to the author whether (70) is true in general for

(a) $s \neq$ integer $+ \frac{1}{p}$; $1 < q < \infty$ or

(b) s arbitrary, $q = 1$.

For $p = q$ the problem is solved by Magenes [35] and Grisvard [22]. For $s \neq$ integer $+ \frac{1}{p}$ and $p = q$ (70) holds. Hence it follows

$$(\overset{o}{W}{}_p^{m_1}(\Omega), \overset{o}{W}{}_p^{m_2}(\Omega))_{\theta,p} = \overset{o}{B}{}^s_{p,p}(\Omega) ; \tag{71}$$

$s \neq$ integer $+ \frac{1}{p}$; $1 < p < \infty$.

We formulate a problem which is closely connected with (70). We consider the operator

$$(Sf)(x) = f(x) \quad \text{for} \quad x \in \Omega ,$$
$$= 0 \quad \text{for} \quad x \in R_n - \Omega .$$

It is not difficult to prove that S is a continuous operator from $\overset{o}{B}{}^s_{p,q}(\Omega)$ into $B^s_{p,q}(R_n)$ iff (70) holds. Operators of such type are considered by Lions-Magenes [30, 33, 35] (especially [35], Theorem 6.2).

Finally we formulate a result due to Grisvard [22]. Let Ω be a bounded domain with $\partial\Omega \in C^\infty$. We consider the differential operators B_j ,

$$(B_j f)(x) = \sum_{|\beta| \le m_j} b_{j\beta}(x) D^\beta f , \quad b_{j\beta}(x) \in C^\infty(\overline{\Omega}) ,$$

$j = 1, 2, \ldots, m$ with

$$0 \le m_1 < m_2 < \ldots < m_m < r ,$$

and

$$b_j(x,\nu_x) \neq 0 \quad \text{for} \quad j = 1, \ldots, m \quad \text{and} \quad x \in \partial\Omega .$$

Then

$$b_j(x,\xi) = \sum_{|\beta|=m_j} b_{j\beta}(x)\xi^\beta$$

with

$$\xi = (\xi_1,\ldots,\xi_n) \in R_n \quad \text{and} \quad \xi^\beta = \xi_1^{\beta_1}\ldots\xi_n^{\beta_n} ,$$

r being an integer. We define

$$W^r_{p,\{B_j\}}(\Omega) = \{f \mid f \in W^r_p(\Omega) , \; B_j f|_{\partial\Omega} = 0 \; ; \; j = 1, \ldots, m\} .$$

Theorem 18. Let Ω be a bounded domain with $\partial\Omega \in C^\infty$.

(a) Let be $1 < p < \infty$; $0 < \theta < 1$ and $r(1 - \theta) \neq$ integer.

Then

$$(W^r_{p,\{B_j\}}(\Omega), L_p(\Omega))_{\theta,p} = \Big\{f \mid f \in W_p^{r(1-\theta)}(\Omega) ,$$

$$B_j f|_{\partial\Omega} = 0 \quad \text{for} \quad m_j < r(1 - \theta) - \frac{1}{p} ,$$

$$\int_\Omega \frac{1}{d(x)} |B_j f(x)|^p \, dx < \infty \quad \text{for} \quad m_j = r(1 - \theta) - \frac{1}{p}\Big\} .$$

Here $d(x)$ is the distance of $x \in \Omega$ from $\partial\Omega$.

(b) For $p = 2$ the result (a) is true also for $r(1 - \theta) =$ integer.

The theorem is due to Grisvard [22] (Theorem 8.1, Theorem 8.1´) and follows from Grisvard´s result and

$$(H_0,H_1)_{\theta,2} = [H_0,H_1]_\theta$$

for two Hilbert spaces $H_0 \subset H_1$.

1.4. Interpolation theory for function spaces with weights I

1.4.1. Definitions. We consider a special case of function spaces with weights defined in (bounded or unbounded) domains. We have the same two kinds of domains as described in 1.3.1: Unbounded domains with the cone property and bounded domains with $\partial\Omega \in C^\infty$. We start with a description of the considered weight functions.

(A) Let Ω be an unbounded domain with the cone property. We denote the cone by K. Ω^λ has the same meaning as in (65). We consider a weight function $\sigma(x)$ with:

(72) $$\sigma(x) > 0 \quad \text{for } x \in \Omega, \quad \sigma(x) \text{ continuous in } \Omega,$$

there exist numbers λ, C and c; $\lambda > 0$; $C > 0$; $c > 0$ with

(73) $$\sigma(x) \leq c \min_{|y-x|\leq\lambda} \sigma(y) \quad \text{for } x \in \Omega - \Omega^{2\lambda}$$

(74) $$\sigma(z + \lambda_1 y) \leq \sigma(z + \lambda_2 y)$$

$$\text{for } z \in \partial\Omega;\ y \in K,\ |y| \leq 1,\ 0 < \lambda_1 \leq \lambda_2 \leq C\lambda,$$

where

$$\{x \mid x = z + \mu y;\ z \in \partial\Omega;\ y \in K,\ |y| \leq 1;\ 0 < \mu < C\} \supset \Omega^{2\lambda}.$$

(B) Let Ω be an unbounded domain with the cone property. We consider another class of weight functions $\sigma(x)$ with (72), (73) and

(74′) $$\sigma(z + \lambda_1 y) \geq \sigma(z + \lambda_2 y)$$

$$\text{for } z \in \partial\Omega,\ y \in K,\ |y| \leq 1,\ 0 < \lambda_1 \leq \lambda_2 \leq C\lambda.$$

(C) Let Ω be a bounded domain with $\partial\Omega \in C^\infty$. Let c_1, c_2, λ be positive numbers and $\varrho(t)$ a continuous function in $(0,\lambda)$ with

$$\varrho(t) > 0 \quad \text{for } 0 < t < \lambda,$$

$$\varrho(t_1) \leq \varrho(t_2) \quad \text{for } 0 < t_1 \leq t_2 < \lambda.$$

Then we consider a weight function $\sigma(x)$ with (72) and

(75) $$c_1 \varrho(d(x)) \leq \sigma(x) \leq c_2 \varrho(d(x)), \quad x \in \Omega^\lambda,$$

$d(x)$ is the distance of $x \in \Omega$ from $\partial\Omega$.

(D) Let Ω be a bounded domain with $\partial\Omega \in C^\infty$. Let c_1, c_2, λ be positive numbers and $\varrho(t)$ a continuous function in $(0,\lambda)$ with

$$\varrho(t) > 0 \quad \text{for} \quad 0 < t < \lambda ,$$

$$\varrho(t_1) \geqq \varrho(t_2) \quad \text{for} \quad 0 < t_1 \leqq t_2 < \lambda .$$

Then we consider a weight function $\sigma(x)$ with (72) and (75).

The assumptions on the weight function $\sigma(x)$ are rather weak. (73) is a weak smoothness property. (74) ((74´)) is a decreasing (increasing) property near the boundary. The same holds for (75).

Let Ω be either an unbounded domain with the cone property or a bounded domain with $\partial\Omega \in C^\infty$. Let m be an integer; $m = 0, 1, 2, \dots$; $1 < p < \infty$ and $\sigma(x)$ a weight function of the type (A), (B), (C) or (D). We write

$$(76) \qquad W^m_{p,\sigma}(\Omega) = \Big\{ f \mid f \in D'(\Omega) ,$$

$$\|f\|_{W^m_{p,\sigma}} = \sum_{|\alpha|\leqq m} \left(\int_\Omega \sigma(x) |D^\alpha f(x)|^p \, dx \right)^{\frac{1}{p}} < \infty \Big\} .$$

$D'(\Omega)$ is the set of distributions in Ω. Further we denote by $\overset{\circ}{W}{}^m_{p,\sigma}(\Omega)$ the completion of $C_0^\infty(\Omega)$ in $W^m_{p,\sigma}(\Omega)$. We write $L_{p,\sigma}(\Omega) = W^0_{p,\sigma}(\Omega)$.

We describe some special cases. Obviously $\sigma(x) \equiv 1$ leads to the Sobolev spaces.

Let be $\Omega = R_n$ or $\Omega = R_n^+ = \{x \mid x_n > 0\}$ and $\sigma(x) = (1 + |x|)^\beta$; $\beta < 0$. Then $W^m_{p,\sigma}(\Omega)$ is similar to the spaces of Kudrjavcev [27]. For a generalization see Ju. S. Nikol´skij [40].

Perhaps the most interesting case is: Ω bounded or unbounded domain,

$$\sigma(x) = d^\varkappa(x) .$$

Spaces of such or similar type have been considered by many authors. For our purpose the papers of Grisvard [20] and Uspenskij [65] are important. We refer also to [11] and [63]. Spaces of such type are also interesting for boundary value problems for regular elliptic differential equations, see [39] and [18].

It is not difficult to see that the spaces $\Xi^m(\Omega)$ introduced by Lions-Magenes in [33], Chapter 2, 6.3, are special cases of the spaces $W^m_{p,\sigma}(\Omega)$,

$$\Xi^m(\Omega) = W^m_{2,d^{2m}(x)}(\Omega) .$$

The spaces $W^m_{p,d^{pm}(x)}(\Omega)$ are the generalizations of the spaces $\Xi^m(\Omega)$ mentioned in [33], Chapter 1, Problem 18.4 and Chapter 2, Problem 11.3.

1.4.2. Embedding theorems and extension theorems, the spaces $\overset{\circ}{W}{}^m_{p,d^\varkappa}$. We start with a general result. If Ω is a domain in R_n we denote by $C^\infty(\overline{\Omega})$ the set of all complex infinitely differentiable functions $f(x)$ vanishing for large $|x|$ whose derivatives are continuous in $\overline{\Omega}$.

Theorem 19. In the cases (A) and (C) of 1.4.1 $C^\infty(\overline{\Omega})$ is dense in $W^m_{p,\sigma}(\Omega)$.

It is not known to the author whether for the (rather wide) class of spaces defined in 1.4.1, final embedding theorems and extension theorems in the sense of Theorems 11 and 12 have been proved. Nevertheless, for the interesting case $\sigma(x) = d^\varkappa(x)$ Uspenskij proved a result in this direction. For a bounded domain with $\partial\Omega \in C^\infty$ we denote by $\nu = \nu_y$ the inner normal for $y \in \partial\Omega$.

Theorem 20. Let Ω be a bounded domain with $\partial\Omega \in C^\infty$. Let m be an integer; $m = 1, 2, \ldots$ and $1 < p < \infty$. Let be $\sigma(x) = d^\varkappa(x)$ with $mp - 1 > \varkappa \geq 0$. Then

$$Kf = \left\{ f, \frac{\partial f}{\partial \nu}, \ldots, \frac{\partial^{[m-\varkappa/p-1/p]^-} f}{\partial \nu^{[m-\varkappa/p-1/p]^-}} \right\} \tag{77}$$

is a linear and continuous map from $W^m_{p,\sigma}(\Omega)$ onto

$$\prod_{j=0}^{[m-\varkappa/p-1/p]^-} B^{m-\varkappa/p-1/p-j}_{p,p}(\partial\Omega) . \tag{78}$$

This is an embedding theorem and extension theorem in the sense of Theorem 11. A proof is given by Uspenskij [65] who considered also more general (anisotropic) cases. (We remark that our definition and Uspenskij's definition are different. But with the help of Uspenskij's paper it is not difficult to show that our spaces and Uspenskij's spaces are the same and the norms are equivalent.) See also Grisvard [20], Theorem 8.1.

Finally we formulate an analogous theorem to Theorem 12.

Theorem 21. Let Ω be a bounded domain with $\partial\Omega \in C^\infty$. Let m be an integer; $m = 1, 2, \ldots$ and $1 < p < \infty$. Let be $\sigma(x) = d^{\varkappa}(x)$. For $mp - 1 > \varkappa \geq 0$ it is

$$\mathring{W}^m_{p,\sigma}(\Omega) = \left\{ f \mid f \in W^m_{p,\sigma}(\Omega) \, , \; \frac{\partial^j f}{\partial \nu^j}\bigg|_{\partial\Omega} = 0 \text{ for } j = 0, \ldots, [m - \frac{\varkappa}{p} - \frac{1}{p}]^- \right\} . \tag{79}$$

For $\infty > \varkappa \geq mp - 1$ it is

$$\mathring{W}^m_{p,\sigma}(\Omega) = W^m_{p,\sigma}(\Omega) . \tag{80}$$

The theorem follows from Theorem 2 of [63]. We refer also to the papers of Besov-Kadlec-Kufner [11] and of Geymonat-Grisvard [18].

1.4.3. Interpolation theory. First we formulate a partial generalization of Theorem 16 and Theorem 9.

Theorem 22. Let Ω be an unbounded domain with the cone property and let K be a cone with (33). Let $\sigma(x)$ be a wieght function of type (A), Section 1.4.1. Let m_1 and m_2 be integers; $0 \leq m_1 < m_2 < \infty$; $1 < p < \infty$; $0 < \theta < 1$; $1 \leq q \leq \infty$; $s = (1 - \theta)m_1 + \theta m_2$. Then

$$B^s_{p,q,\sigma}(\Omega) = (W^{m_1}_{p,\sigma}(\Omega), W^{m_2}_{p,\sigma}(\Omega))_{\theta,q} = \{ f \mid f \in L_{p,\sigma}(\Omega) \, , \; \|f\|^{(r)} < \infty \} \; ; \quad r = 1, 2, \tag{81}$$

where $\|f\|^{(r)}$ are the norms described in (37) and (38) where we have to replace $L_p(\Omega)$ by $L_{p,\sigma}(\Omega)$ and K by $K \cap \{h \mid |h| \leq 1\}$. For j and ℓ we have the same conditions as in Theorem 9: j and ℓ are integers; $0 \leq j < s$; $\ell > s - j$. The norms $\|f\|^{(r)}$ are equivalent to the norm of the space $B^s_{p,q,\sigma}(\Omega)$.

This theorem is proved in [64]. $B^s_{p,q,\sigma}(\Omega)$ are Besov spaces with weights. Perhaps the result is also interesting from the point of view of equivalent norms. The assumptions for the weights $\sigma(x)$ are rather weak. So it seems to be somewhat complicated to show in a direct way that the mentioned norms are equivalent to each other. It would be interesting to prove that the norm of the interpolation space $B^s_{p,q,\sigma}(\Omega)$ is also equivalent to the norm analogous to (39). Perhaps it is true, but the method developed in [64] is not strong enough to obtain such a result.

Next we formulate an analogous result for bounded domains with the smooth boundary. We use the definition of Ω_h given in (41).

Theorem 23. Let Ω be a bounded domain with $\partial\Omega \in C^{\infty}$. Let $\sigma(x)$ be a weight function of type (C), Section 1.4.1. Let m_1 and m_2 be two integers; $0 \leq m_1 < m_2 < \infty$; $1 < p < \infty$; $0 < \theta < 1$; $1 \leq q \leq \infty$; $s = (1-\theta)m_1 + \theta m_2$. Then

$$(82)\qquad B^s_{p,q,\sigma}(\Omega) = (W^{m_1}_{p,\sigma}(\Omega), W^{m_2}_{p,\sigma}(\Omega))_{\theta,q} = \{f \mid f \in L_{p,\sigma}(\Omega)\ ,\ \|f\|^{(r)} < \infty\}\ ;\quad r = 1, 2$$

where $\|f\|^{(r)}$ are the norms described in (42) and (43), where we have to replace $L_p(\Omega_h)$ by $L_{p,\sigma}(\Omega_h)$. For j and ℓ we have the same conditions as in Theorem 10: j and ℓ are integers; $0 \leq j < s$; $\ell > s - j$. Then the norms $\|f\|^{(r)}$ are equivalent to the norm of the space $B^s_{p,q,\sigma}(\Omega)$.

This theorem is proved in [64]. Special cases are interesting spaces with $\sigma(x) = d^{\varkappa}(x)$ for which Theorem 20 and formula (80) hold. We also obtain many equivalent norms for the interpolation spaces $B^s_{p,q,\sigma}(\Omega)$ (Besov spaces with weights). Of course it would be interesting to show that the norm analogous to (39) is also an equivalent norm in $B^s_{p,q,\sigma}(\Omega)$.

We mention a conclusion of the last theorem, Theorem 20, and a theorem of Grisvard [20] (Theorem 8.1). Let be $\sigma(x) = d^{\varkappa}(x)$; $s > 0$; $1 < p < \infty$; $1 \leq q \leq \infty$; $sp - 1 > \varkappa \geq 0$. Then the operator K defined in (77) is a linear and continuous map from $B^s_{p,q,\sigma}(\Omega)$ into

$$(83)\qquad \prod_{j=0}^{[s-\varkappa/p-1/p]^-} B^{s-\varkappa/p-1/p-j}_{p,q}(\partial\Omega)\ .$$

Now we pass on to interpolation theorems for the spaces $\overset{\circ}{W}{}^m_{p,\sigma}(\Omega)$. We use the definition of Ω^{λ} given in (65). We set

$$\|f\|_{(s,p,q,\sigma)} = \left[\int_0^{\infty} t^{-sq}\|f\|^q_{L_{p,\sigma}(\Omega^t)}\frac{dt}{t}\right]^{\frac{1}{q}}\ .$$

Further we write for an unbounded domain with the cone property and for a weight function of type (B), Section 1.4.1

$$\|f\|^*_{B^s_{p,q,\sigma}} = \left[\int_{K\cap\{y \mid |y|\leq 1\}} |y|^{-sq}\left(\int_{\Omega} |(\Delta^{\ell}_y f)(x)|^p \sigma(x+\ell y)\,dx\right)^{\frac{q}{p}} \frac{dy}{|y|^n}\right]^{\frac{1}{q}}$$

and for a bounded domain, $\partial\Omega \in C^\infty$, and for a weight function of type (D)

(84) $$\|f\|^*_{B^s_{p,q,\sigma}} = \left[\int_{|y|\le\delta} |y|^{-sq}\left(\int_{\Omega_y} |(\Delta^\ell_y f)(x)|^p \sigma(x+\ell y)\, dx\right)^{\frac{q}{p}} \frac{dy}{|y|^n}\right]^{\frac{1}{q}} .$$

Here $\delta > 0$; $1 < p < \infty$; $1 \le q \le \infty$; $s > 0$; integer; $\ell > s$. (For $q = \infty$ we have to modify in the usual way.)

Theorem 24. Let Ω be an unbounded domain with the cone property or a bounded domain with $\partial\Omega \in C^\infty$. Let $\sigma(x)$ be a weight function of type (B), Section 1.4.1 if the domain is unbounded, and of type (D) if the domain is bounded. Let m_1 and m_2 be two integers; $0 \le m_1 < m_2 < \infty$; $1 < p < \infty$; $0 < \Theta < 1$; $1 \le q \le \infty$; $s = (1-\Theta)m_1 + \Theta m_2$. Then

$$(\overset{o}{W}{}^{m_1}_{p,\sigma}(\Omega), \overset{o}{W}{}^{m_2}_{p,\sigma}(\Omega))_{\Theta,q} = \{f \mid f \in L_{p,\sigma}(\Omega) ,\ \|f\|^{*(r)} < \infty\} , \quad r = 1, 2$$

with

$$\|f\|^{*(1)} = \|f\|_{L_{p,\sigma}} + \sum_{j=1}^{n} \left[\left\|\frac{\partial^k u}{\partial x_j^k}\right\|^*_{B^{s-k}_{p,q,\sigma}} + \left\|\frac{\partial^k u}{\partial x_j^k}\right\|_{(s-k,p,q,\sigma)}\right] ,$$

$$\|f\|^{*(2)} = \|f\|_{L_{p,\sigma}} + \sum_{|\alpha|\le k} \left[\|D^\alpha u\|_{B^{s-k}_{p,q,\sigma}} + \|D^\alpha u\|_{(s-k,p,q,\sigma)}\right] .$$

δ in (84) is sufficiently small. k and ℓ are integers ; $0 \le k < s$; $\ell > s - k$. All the norms are equivalent to the norm of $(\overset{o}{W}{}^{m_1}_{p,\sigma}(\Omega), \overset{o}{W}{}^{m_2}_{p,\sigma}(\Omega))_{\Theta,q}$.

The theorem is proved in [64]. It is a generalization of Theorem 17.

We discuss a special interesting case. Let be $\sigma(x) = d^{-\varkappa}(x)$; $\varkappa \ge 0$ and $p = q$. Then

$$\|f\|_{(s,p,p,d^{-\varkappa})} = \left(\int_\Omega d^{-\varkappa-sp}(x)\,|f(x)|^p\, dx\right)^{\frac{1}{p}} .$$

This simplifies the last theorem.

For further interpolation results for $\Omega = R_n^+$, $\sigma(x) = x_n^{\varkappa}$, $m_1 = 0$, $m_2 = 1$, we refer to Grisvard's paper [20].

1.5. Interpolation theory for function spaces with weights II

1.5.1. Definitions. In 1.4 we considered a special class of function spaces with weights. But for applications to singular elliptic differential operators we need more general function spaces with weights. On one side we generalize the type of the spaces, on the other side we restrict our attention to the one-dimensional case and to the case of Hilbert spaces.

Let be $-\infty < a < b < \infty$. In the closed interval $[a,b]$ we consider the weight function $p(x)$ with

$$p(x) \in C^{\infty}([a,b]) \,, \quad p(x) > 0 \quad \text{for} \quad x \in (a,b) \,, \tag{85}$$

$$\infty > \lim_{x \downarrow a} \frac{p(x)}{x - a} = C_a > 0 \,, \quad \infty > \lim_{x \uparrow b} \frac{p(x)}{b - x} = C_b > 0 \,.$$

We define the spaces $H^{\mu}_{\varkappa_1,\varkappa_2} = H^{\mu}_{\varkappa_1,\varkappa_2}([a,b])$.

1. Let μ be an integer; $\mu = 1, 2, \ldots$. Let $\varkappa_1$ and $\varkappa_2$ be two real numbers. We set

$$H^{\mu}_{\varkappa_1,\varkappa_2} = \left\{ f \mid f \in D'(a,b) \,, \; \|f\|_{H^{\mu}_{\varkappa_1,\varkappa_2}} = \right. \tag{86a}$$

$$\left. = \left(\int_a^b \left(p^{\varkappa_1}(x)\,|f^{(\mu)}(x)|^2 + p^{\varkappa_2}(x)\,|f(x)|^2 \right) dx \right)^{\frac{1}{2}} < \infty \right\} .$$

(We remind that $D'(a,b)$ denotes the set of distributions in (a,b) .) If $\mu = 0$ we assume $\varkappa_1 = \varkappa_2$ and set

$$H^0_{\varkappa_1,\varkappa_2} = L_{2,\varkappa_1} \,.$$

2. Let μ be a positive number with $\mu = [\mu] + \{\mu\}$; $[\mu]$ integer; $0 < \{\mu\} < 1$. We set

$$H^{\mu}_{\varkappa_1,\varkappa_2} = \left\{ f \mid f \in D'(a,b) \,, \; \|f\|_{H^{\mu}_{\varkappa_1,\varkappa_2}} = \right. \tag{86b}$$

$$= \left(\int_a^b \int_a^b \frac{|p^{\varkappa_1/2}(x) f^{([\mu])}(x) - p^{\varkappa_1/2}(y) f^{([\mu])}(y)|^2}{|x - y|^{1+2\{\mu\}}} dx\, dy + \right.$$

$$\left. + \int_a^b p^{\varkappa_2}(x) |f(x)|^2 \, dx \right)^{\frac{1}{2}} < \infty \Big\} .$$

By $\overset{o}{H}{}^{\mu}_{\varkappa_1,\varkappa_2}$ we denote the completion of $C_0^{\infty}((a,b))$ in $H^{\mu}_{\varkappa_1,\varkappa_2}$. These spaces are introduced in [59, 60].

1.5.2. Interpolation theory. It is not difficult to see that the spaces $H^{\mu}_{\varkappa_1,\varkappa_2}$ are Hilbert spaces. The interpolation theory for these spaces is developed in [60]. We formulate some results.

Theorem 25. (a) Let be $\mu \geqq 0$; $\varkappa_1$ and $\varkappa_2$ real numbers; $\varkappa_1 \geqq \varkappa_2 + 2\mu$. Then $C_0^{\infty}((a,b))$ is a dense subset in $H^{\mu}_{\varkappa_1,\varkappa_2}$.

(b) Let be $\mu \geqq 0$; $\nu \geqq 0$; $\mu \neq \nu$; $\varkappa_1$, $\varkappa_2$, ϱ_1 , ϱ_2 real numbers; $\varkappa_1 \geqq \varkappa_2 + 2\mu$; $\varrho_1 \geqq \varrho_2 + 2\nu$; and $(\varkappa_1 - \varkappa_2)\nu = (\varrho_1 - \varrho_2)\mu$. Then

$$(H^{\mu}_{\varkappa_1,\varkappa_2}, H^{\nu}_{\varrho_1,\varrho_2})_{\Theta,2} = H^{\eta}_{\sigma_1,\sigma_2} \tag{87}$$

with

$$\eta = (1 - \Theta)\mu + \Theta\nu \; ; \quad \sigma_i = (1 - \Theta)\varkappa_i + \Theta\varrho_i \; ;$$
$$i = 1, 2 \; ; \quad 0 < \Theta < 1 \, .$$

(c) (87) is also true for $\mu = \nu =$ integer and the other assumptions.

The interpolation theory for the spaces $H^{\mu}_{\varkappa_1,\varkappa_2}$ with $\varkappa_1 < \varkappa_2 + 2\mu$ is more complicated.

Theorem 26. Let be $\mu \geqq 0$; $\nu \geqq 0$; $\mu \neq \nu$; $\mu \neq \frac{1}{2}, \frac{3}{2}, \frac{5}{2}, \ldots$; $\nu \neq \frac{1}{2}, \frac{3}{2}, \frac{5}{2}, \ldots$; $\varkappa_1$, $\varkappa_2$, ϱ_1 , ϱ_2 real numbers with $\varkappa_1 \leqq \varkappa_2 + 2\mu$ and $\varrho_1 \leqq \varrho_2 + 2\nu$. Further we assume $\varkappa_1 - 2\mu \neq -1, -3, \ldots, 1 - 2[\mu]$ and $\varrho_1 - 2\nu \neq -1, -3, \ldots, 1 - 2[\nu]$. Then

$$(\overset{o}{H}{}^{\mu}_{\varkappa_1,\varkappa_2}, \overset{o}{H}{}^{\nu}_{\varrho_1,\varrho_2})_{\Theta,2} = H^{\eta}_{\sigma,\sigma-2\eta} = \overset{o}{H}{}^{\eta}_{\sigma,\sigma-2\eta} \tag{88}$$

with

$$\eta = \mu(1 - \Theta) + \nu\Theta \; ; \quad \sigma = \varkappa_1(1 - \Theta) + \varrho_1\Theta \, .$$

The situation is similar to the situation for function spaces without weights where we had "singular" values for the indices. For details we refer to [60]. We have to do a remark. In [60] we formulated the result without the assumption $\mu \neq \frac{1}{2}, \frac{3}{2}, \frac{5}{2}, \dots$; $\nu \neq \frac{1}{2}, \frac{3}{2}, \frac{5}{2}, \dots$. But we must correct our result in the described way. The last theorem is also true for $\mu = \nu =$ integer and the other assumptions.

The last two theorems show that the spaces $H^{\mu}_{\varkappa_1,\varkappa_2}$ and $\overset{\circ}{H}{}^{\mu}_{\varkappa_1,\varkappa_2}$ have different interpolation properties. It is possible to prove that

$$(H^2_{2,0}, L_2)_{1/2,2} = H^1_{1,0} .$$

On the other hand, it follows from (88) that

$$(\overset{\circ}{H}{}^2_{2,0}, L_2)_{1/2,2} = H^1_{1,-1} .$$

$H^1_{1,-1}$ is a dense subset of $H^1_{1,0}$ but the norms of the two spaces are not equivalent [60].

Perhaps it will be possible to transfer some results to the more general Banach space case and to give a more systematic treatment of the interpolation theory of these spaces.

1.6. Interpolation theory for function spaces with weights III

1.6.1. Definitions. Now we consider function spaces with strong singularities near the boundary. We restrict our attention to the Hilbert space case. But the method developed in [57] seems to be applicable also to the Banach space case.

Let Ω be an arbitrary domain in R_n (bounded or unbounded, without any smoothness assumptions). We consider two kinds of weight functions.

A. Let $p(x)$ be a differentiable function in Ω with:

1. $p(x) > 0$ for $x \in \Omega$;
2. there exist $C \geqq 0$ and $\sigma \geqq 0$ with

$$|\nabla p(x)| \leqq Cp^{1+\sigma}(x) ; \tag{89}$$

3. for every positive number K there exist two numbers ε_K and N_K with

$$p(x) \geqq K \text{ for } d(x) \leqq \varepsilon_K \text{ or } |x| \geqq N_K , \quad (x \in \Omega) .$$

(We remind that $d(x)$ denotes the distance of a point $x \in \Omega$ from the boundary $\partial\Omega$.)

B. Let $p(x)$ be a continuous function in Ω with A1, A3 and there exist $c > 0$ and $\sigma \geq 0$ with

$$\inf_{\substack{p(x)=2^{k-1} \\ p(y)=2^k}} |x - y| \geq c\, 2^{-k\sigma} \; ; \quad k = k_0, k_0 + 1, \dots . \tag{90}$$

It is possible to prove that each weight function of type A is also a weight function of type B, especially (90) is true with the same σ , see [57].

It is easy to give examples of such weight functions. For $\Omega = R_n$ each "slowly increasing" differentiable positive function is a function of such type (for instance a function increasing like a polynomial or an e-function). For a bounded domain,

$$p(x) = d^{-1/\sigma}(x)$$

is a function of type B. If the boundary of the bounded domain is smooth then we can construct a positive differentiable function $\delta(x)$ with

$$\delta(x) = d(x) \quad \text{near the boundary.}$$

$\delta^{-1/\sigma}(x)$ is a weight function of type A.

Let m be an integer; $m = 1, 2, \dots$ and $p(x)$ a weight function of type A or of type B. We set

$$W^m_{2,p}(\Omega) = \Big\{ f \mid f \in D'(\Omega) , \tag{91}$$

$$\|f\|_{W^m_{2,p}} = \Big[\int_\Omega \Big(\sum_{|\alpha|=m} |D^\alpha f(x)|^2 + p(x) |f(x)|^2 \Big) dx \Big]^{\frac{1}{2}} < \infty \Big\} .$$

Evidently we can define similar spaces for the Banach space case. The interpolation theory developed in [57] seems to be strong enough also for the general Banach space case. Nevertheless, we restrict our attention to the Hilbert space case.

1.6.2. Interpolation theory.

Theorem 27. Let m be an integer; $m = 1, 2, \dots$. Let be $0 \leq \varkappa \leq 1$; $0 < \theta < 1$ and let $p(x)$ be a weight function of type A in an arbitrary domain $\Omega \subset R_n$ with $0 \leq \sigma \leq 1/(2m)$ in (89). Then

$C_0^\infty(\Omega)$ is dense in $W^m_{2,p}(\Omega)$ and

$$(W^m_{2,p}(\Omega), L_{2,p^\varkappa}(\Omega))_{\theta,2}$$

is the completion of $C_0^\infty(\Omega)$ in the norm

$$(92) \qquad \Big(\sum_{|\alpha|=[(1-\theta)m]} \| p^{\varkappa\theta/2} D^\alpha v \|^2_{W_2^{\{(1-\theta)m\}}(\Omega)} + \\ + \| p^{(\varkappa\theta+1-\theta)/2} v \|^2_{L_2(\Omega)} \Big)^{\frac{1}{2}} .$$

(We remind that $s = [s] + \{s\}$ with $[s]$ integer, $0 \leq \{s\} < 1$.)

For the special case $\varkappa = 0$ we can weaken the assumptions and formulate the following

Theorem 28. Let m be an integer; $m = 1, 2, \ldots$. Let be $0 < < \theta < 1$, and let $p(x)$ be a weight function of type B in an arbitrary domain $\Omega \subset R_n$ with $0 \leq \sigma \leq 1/(2m)$ in (90). Then $C_0^\infty(\Omega)$ is dense in $W^m_{2,p}(\Omega)$ and

$$(W^m_{2,p}(\Omega), L_2(\Omega))_{\theta,2}$$

is the completion of $C_0^\infty(\Omega)$ in the norm

$$(93) \qquad \Big(\|v\|^2_{W_2^{(1-\theta)m}(\Omega)} + \| p^{(1-\theta)/2}\, v \|^2_{L_2(\Omega)} \Big)^{\frac{1}{2}} .$$

As a special case of the last theorem we can consider a bounded domain (without any smoothness properties) and the weight function $p(x) = d^{-\tau}(x)$ with $\tau \geq 2m$. We formulate a more general result. We denote by $\mathring{W}^m_{2,p}(\Omega)$ the completion of $C_0^\infty(\Omega)$ in the space $W^m_{2,p}(\Omega)$.

Theorem 29. Let Ω be a bounded domain with $\partial\Omega \in C^2$. Let $\delta(x)$ be a positive differentiable function with $\delta(x) = d(x)$ near the boundary. Let be $0 \leq \tau < \infty$ and $0 \leq \varkappa \leq 1$. Then

$$(\mathring{W}^m_{2,d^{-\tau}}(\Omega), L_{2,d^{-\varkappa\tau}}(\Omega))_{\theta,2}$$

is the completion of $C_0^\infty(\Omega)$ in the norm

$$(94) \qquad \Big(\sum_{|\alpha| = [(1-\Theta)m]} \| \delta^{-\tau\varkappa\Theta/2}(x) D^\alpha v(x) \|^2_{W_2^{\{(1-\Theta)m\}}(\Omega)} + \\ + \| d^{-\tau(\varkappa\Theta + 1-\Theta)/2}(x) v(x) \|^2_{L_2(\Omega)} \Big)^{\frac{1}{2}} .$$

For $\tau \geqq 2m$ the theorem is a special case of Theorem 27. For $0 \leqq \tau < 2m$ we need additional considerations. The last theorem shows that the result is independent of the choice of $\delta(x)$. All theorems of this section are proved in [57].

2. ELLIPTIC DIFFERENTIAL OPERATORS

2.1. Regular elliptic differential operators

2.1.1. Definitions. In this section we describe regular elliptic boundary value problems. The basis of our study are the papers of Agmon-Douglis-Nirenberg [4], Agmon [1, 2] and Lions-Magenes [29, 30, 31, 32, 33]. Further we refer to Berezanskij [8], Geymonat-Grisvard [19] and Magenes [35]. We start with the necessary definitions.

Let Ω be a bounded domain in R_n with the smooth boundary $\partial\Omega \in C^\infty$. We consider a differential operator A of even order,

$$(95) \qquad Au = \sum_{|\alpha| \leqq 2m} a_\alpha(x) D^\alpha u , \quad a_\alpha(x) \in C^\infty(\overline{\Omega}) .$$

We assume that A is properly elliptic. That means:

1. For all $\xi \in R_n$; $\xi \neq 0$ and all $x \in \overline{\Omega}$ it is

$$(96) \qquad a(x,\xi) = \sum_{|\alpha| = 2m} a_\alpha(x) \xi^\alpha \neq 0 , \quad \xi^\alpha = \xi_1^{\alpha_1} \dots \xi_n^{\alpha_n} .$$

2. For fixed linearly independent $\xi \in R_n$ and $\eta \in R_n$, $a(x,\xi + \tau\eta)$ has (as a polynomial in τ) exactly m roots τ_j^+ ; $j = 1, \dots, m$ with $\operatorname{Im} \tau_j^+ > 0$, and m roots τ_j^- ; $j = 1, \dots, m$ with $\operatorname{Im} \tau_j^- < 0$. We set

$$(97) \qquad a^+_{\xi,\eta,x}(\tau) = \prod_{j=1}^{m} (\tau - \tau_j^+) .$$

We remark that for $n \geq 3$ the latter condition follows from the former.

Besides the operator A we consider boundary operators B_j ; $j = 1, \dots, m$,

(98) $$B_j u = \sum_{|\beta| \leq m_j} b_{j\beta}(x) D^\beta u \ , \quad b_{j\beta}(x) \in C^\infty(\partial\Omega) \ .$$

We assume that the operators B_j are normal. That means:

3. $0 \leq m_1 < m_2 < \dots < m_m < 2m - 1$ for $x \in \partial\Omega$ and the normal vector $\nu = \nu_x$ in x is

(99) $$b_j(x, \nu_x) \neq 0 \ ,$$

where

(100) $$\dot{b}_j(x, \xi) = \sum_{|\beta| = m_j} b_{j\beta}(x) \xi^\beta \ .$$

Further we assume the complementing condition.

4. Let be $x \in \partial\Omega$, $\xi \neq 0$ a tangential vector at x . Then

(101) the polynomials in τ $b_j(x, \xi + \tau\nu_x)$ are linearly independent modulo $a^+_{\xi,\nu,x}(\tau)$.

These are the well-known conditions for a regular elliptic boundary value problem. Next we describe conditions which are somewhat stricter.

Let A be a proper elliptic differential operator of type (95). That means that 1 and 2 hold. Further we have m boundary operators B_j which are normal. That means that 3 holds. We replace 1 and 4 by stricter conditions 1′ and 4′:

1′. It is $a(x, \xi) \neq 0$ for $x \in \bar{\Omega}$ and $\xi \in R_n$, $\xi \neq 0$. Further, there exists a number Θ with $-\pi < \Theta \leq \pi$ and

(102) $$(-1)^m \frac{a(x, \xi)}{|a(x, \xi)|} \neq e^{i\Theta} \ ; \quad x \in \bar{\Omega} \ ; \quad \xi \neq 0 \ .$$

4′. Let λ be a number with $\lambda = |\lambda| e^{i\Theta}$ (where Θ has the meaning from (102)) and

(103) $$(-1)^m a(x, \xi + \tau\nu_x) - \lambda = a^+_{\xi,\nu,x,\lambda}(\tau) a^-_{\xi,\nu,x,\lambda}(\tau) \ ,$$

where

$$(104) \qquad a^{+}_{\xi,\nu,x,\lambda}(\tau) = \prod_{j=1}^{m} (\tau - \tau^{+}_{j}(\lambda))$$

is the polynomial with the m roots of the left-hand side of (103) with $\operatorname{Im} \tau^{+}_{j} > 0$. $\xi \neq 0$ is a tangential vector, ν_x the normal vector. We assume that the polynomials $b_j(x,\xi + \tau\nu_x)$ are linearly independent modulo $a^{+}_{\xi,\nu,x,\lambda}(\tau)$ for each λ with $\lambda = |\lambda| e^{i\theta}$.

We call 1′, 2, 3, 4 the Agmon-conditions [1].

<u>2.1.2. Boundary value problems in spaces without weights.</u> We consider the homogeneous boundary value problem

$$(105) \qquad Au = f \ , \quad B_j u|_{\partial\Omega} = 0 \ ; \quad j = 1, \ldots, m \ ;$$

and the non-homogeneous boundary value problem

$$(106) \qquad Au = f \ , \quad B_j u|_{\partial\Omega} = g_j \ ; \quad j = 1, \ldots, m \ ;$$

in Sobolev-Slobodeckij-Besov spaces. We set

$$(107) \qquad \mathfrak{A} = (A, B_1, \ldots, B_m) \ .$$

The basis for the investigations are the a priori estimates and the theory of adjoint boundary value problems. We do not describe here in detail these topics and refer to Agmon-Douglis-Nirenberg [4], Agmon [1], Berezanskij [8], Schechter [47] and Lions-Magenes [32, 33].

Let $W^s_p(\Omega)$ be a Sobolev-Slobodeckij space; $1 < p < \infty$; $s \geq 2m$. We set

$$W^s_{p,\{B_j\}}(\Omega) = \{f \mid f \in W^s_p(\Omega) \ , \ B_j f|_{\partial\Omega} = 0\} \ .$$

In a similar way we define $B^s_{p,q,\{B_j\}}$; $1 < p < \infty$; $1 \leq q \leq \infty$; $s > 2m$. Theorem 11 and $m_j \leq 2m - 1$ show that this definition is meaningful. (We remind that the spaces W^s_p are special cases of the spaces H^s_p and $B^s_{p,p}$.)

On the basis of Theorem 11 it follows immediately that the operator A leads to a linear continuous map from $W^{s+2m}_{p,\{B_j\}}(\Omega)$ $(B^{s+2m}_{p,q,\{B_j\}}(\Omega))$ into $W^s_p(\Omega)$ $(B^s_{p,q}(\Omega))$ and the operator $\mathfrak{A}$ leads to a linear continuous map from $W^{s+2m}_p(\Omega)$ $(B^{s+2m}_{p,q}(\Omega))$ into

$$W_p^s(\Omega) \times \prod_{j=1}^{m} B_{p,p}^{s+2m-m_j-1/p}(\partial\Omega)$$

$$\left(B_{p,q}^s(\Omega) \times \prod_{j=1}^{m} B_{p,q}^{s+2m-m_j-1/p}(\partial\Omega)\right) ;$$

$s \geqq 0$ for the W-spaces; $s > 0$ for the B-spaces; $1 < p < \infty$; $1 \leqq q \leqq \infty$. There is an interesting question whether A or $\mathfrak{A}$ (or more generally $A - \lambda E$ or $\mathfrak{A} - \lambda E$) leads to isomorphic maps. Modifying slightly the usual definition we denote by S_A $(S_{\mathfrak{A}})$ the set of all complex numbers for which the map $A - \lambda E$ $(\mathfrak{A} - \lambda E)$ is not isomorphic from $W_{p,\{B_j\}}^{s+2m}(\Omega)$ or $B_{p,q,\{B_j\}}^{s+2m}(\Omega)$ $(W_p^{s+2m}(\Omega)$ or $B_{p,q}^{s+2m}(\Omega))$ onto $W_p^s(\Omega)$ or $B_{p,q}^s(\Omega)$

$$\left(W_p^s(\Omega) \times \prod_{j=1}^{m} B_{p,p}^{s+2m-m_j-1/p}(\partial\Omega)\right.$$

or

$$\left.B_{p,q}^s(\Omega) \times \prod_{j=1}^{m} B_{p,q}^{s+2m-m_j-1/p}(\partial\Omega)\right) .$$

At the first sight it seems necessary to write $S_{A,s,p,q}$ instead of S_A . However, the following result shows that we can write S_A . We call S_A the spectrum.

Theorem 30. (a) Let $\{A, B_j\}$ be a regular elliptic boundary value problem. Let 1, 2, 3, 4 of 2.1.1 hold. Then $S_A = S_{\mathfrak{A}}$ is independent of the choice of s , p , q ; $s > 0$; $1 < p < \infty$; $1 \leqq q \leqq \infty$, as well as of the choice of the spaces W_p^s or $B_{p,q}^s$ (for the W-spaces we can also set $s = 0$). S_A is either the whole complex plane or S_A is discrete. The operators $A - \lambda E$ and $\mathfrak{A} - \lambda E$ are ϕ-operators.

(b) Let $\{A, B_j\}$ be a regular elliptic boundary value problem with the Agmon-conditions 1′, 2, 3, 4′ of 2.1.1. Then $S_A = S_{\mathfrak{A}}$ is discrete.

First we explain the theorem. A linear continuous operator C acting from the Banach space B_0 into a Banach space B_1 is called a ϕ-operator if

$$\dim \{u \mid Cu = 0\} < \infty ,$$

the range $R(C)$ is closed, and

$$\dim B_1 / R(C) < \infty .$$

S_A discrete means that S_A is an enumerable set of isolated complex numbers, i. e. in any compact set of complex plane there is only a finite number of points $\lambda \in S_A$. If we interpret A as an unbounded operator in $L_p(\Omega)$ with the domain of definition $D(A) = W^{2m}_{p,\{B_j\}}(\Omega)$ then all points $\lambda \in S_A$ are eigenvalues of a finite algebraic multiplicity,

$$\dim \left[\bigcup_{k=1}^{\infty} \{f \mid (A - \lambda E)^k f = 0\} \right] < \infty .$$

For the Sobolev spaces $W^{\ell}_p(\Omega)$; $1 < p < \infty$; $\ell = 1, 2, \ldots$, Theorem 30 (a) follows from the result of Agmon-Douglis-Nirenberg [4], for the Hilbert space case see also Agmon [2], Berezanskij [8], Lions-Magenes [33]. For the Sobolev spaces $W^{\ell}_p(\Omega)$ the part (b) of the last theorem is due to Agmon [1]. The more general cases described in the theorem are due to Lions-Magenes [32, 35], see also Theorem 16 which is the basis for this part of the theorem.

We add some remarks.

1. By means of the adjoint boundary value problem to $\{A, B_j\}$ it is possible to determine more exactly the range of $A - \lambda E$ or $\mathfrak{A} - \lambda E$, see Schechter [47] and Lions-Magenes [32, 33, 35].

2. We assume that the spectrum of A ($\mathfrak{A}$) is discrete. The interesting question of the distribution of the eigenvalues and the completeness of the generalized eigenvectors is considered by many authors. We refer to Browder [12], Agmon [2, 3] and Geymonat-Grisvard [19].

3. We restricted our attention to spaces W^s_p with $s \geqq 0$. One of the main purposes of the paper by Lions-Magenes is an extension of the theory to spaces W^s_p with $s < 0$. We do not describe this interesting theory and refer to Lions-Magenes [29, 30, 31, 32, 33], Magenes [35] and Geymonat-Grisvard [19].

2.1.3. Boundary value problems in spaces with weights. Some authors considered regular elliptic boundary value problems in spaces with weights. For second-order operators we refer to Nečas [39]. Geymonat and Grisvard considered in [18] general regular elliptic boundary value problems of the type described in 2.1.1 in Sobolev spaces with weights. The basis are the spaces $W^k_{p,\sigma}(\Omega)$ with $\sigma(x) = d^{\varkappa}(x)$ described in 1.4.1.

Let $\{A,B_j\}$ be a regular elliptic boundary value problem described in 2.1.1 and 2.1.2. With the weight $\sigma(x) = d^{\varkappa}(x)$; $\varkappa \geqq 0$, the spaces $W^k_{p,\sigma}(\Omega)$ and $B^s_{p,q,\sigma}(\Omega)$ have the meaning of Section 1.4.1 and Section 1.4.3. We remind of Theorem 20 and Theorem 23. For $s \geqq 2m$ and $0 \leqq \varkappa < p - 1$ we set

$$W^s_{p,\sigma,\{B_j\}}(\Omega) = \{f \mid f \in W^s_{p,\sigma}(\Omega) \; , \; B_j u|_{\partial\Omega} = 0 \; ; \; j = 1, \ldots, m\}$$

and similarly $B^s_{p,q,\sigma,\{B_j\}}(\Omega)$ for $s > 2m$. Theorem 20 and formula (83) show that this is meaningful. (We write $W^s_{p,\sigma} = B^s_{p,p,\sigma}$ for $s \neq$ $\neq$ integer .)

The operators A and $\mathfrak{A}$ have the same meaning as in 2.1.2, also the spectrum $S_A = S_{\mathfrak{A}}$.

<u>Theorem 31</u>. Let $\{A,B_j\}$ be a regular elliptic boundary value problem, let 1, 2, 3, 4 of 2.1.1 hold. Let be $\sigma(x) = d^{\varkappa}(x)$ with $0 \leqq \varkappa < p - 1$. Let be $\lambda \notin S_A = S_{\mathfrak{A}}$. Then $A - \lambda E$ is an isomorphic map from $B^{s+2m}_{p,q,\sigma,\{B_j\}}(\Omega)$ onto $B^s_{p,q,\sigma}(\Omega)$ and from $W^{s+2m}_{p,\sigma,\{B_j\}}(\Omega)$ onto $W^s_{p,\sigma}(\Omega)$. $\mathfrak{A} - \lambda E$ is an isomorphic map from $B^{s+2m}_{p,q,\sigma}(\Omega)$ onto

$$B^s_{p,q,\sigma}(\Omega) \times \prod_{j=1}^{m} B^{s+2m-m_j-\varkappa/p-1/p}_{p,q}(\partial\Omega)$$

and from $W^{s+2m}_{p,\sigma}(\Omega)$ onto

$$W^s_{p,\sigma}(\Omega) \times \prod_{j=1}^{m} B^{s+2m-m_j-\varkappa/p-1/p}_{p,p}(\partial\Omega) \; .$$

Here $1 < p < \infty$; $1 \leqq q \leqq \infty$; $s > 0$ for the B-spaces; $s \geqq 0$ for the W-spaces.

For the Sobolev spaces with weights the theorem is due to Geymonat-Grisvard [18]. The general theorem follows from the interpolation theory for these spaces developed in 1.4.3. Geymonat and Grisvard considered also weights $\sigma(x) = d^{\varkappa}(x)$ with $-1 < \varkappa \leqq 0$. For details we refer to [18]. In the general case (not necessarily $\lambda \notin S_A$) the operators $A - \lambda E$ and $\mathfrak{A} - \lambda E$ in the last theorem are Φ-operators.

<u>2.1.4. Fractional powers and Green's functions</u>. We consider two special questions. Let $\{A,B_j\}$ be a regular elliptic boundary value problem. We interpret A as an unbounded operator acting in $L_2(\Omega)$ with the domain of definition

$$D(A) = W^{2m}_{2,\{B_j\}}(\Omega) \; .$$

Theorem 32. Let be $\lambda \notin S_A$. Then

$$(108)\qquad [(A-\lambda E)^{-1}f](x) = \int_\Omega G_\lambda(x,y)f(y)\,dy\ ,\quad f(y) \in L_2(\Omega)\ ,$$

with $G_\lambda(x,y) \in W_2^\varrho(\Omega \times \Omega)$; $\varrho \geq 0$ iff

$$(109)\qquad 2m - \frac{n}{2} > \varrho\ .$$

A proof of this theorem is given in [55]. $G_\lambda(x,y)$ is called Green's function. We add three remarks.

1. The theorem is also true for an arbitrary λ if we consider $A - \lambda E$ as an operator from $W^{2m}_{2,\{B_j\}}(\Omega)/N(A)$ onto $R(A - \lambda E)$ [55].

2. Neither the restriction $\varrho \geq 0$ is necessary. We set $W_2^{-\varrho}(\Omega \times \Omega) = (\mathring{W}_2^\varrho(\Omega \times \Omega))'$; $\varrho \geq 0$. Then (108) and (109) hold in the sense of distributions [55].

3. Local smoothness properties for Green's function $G_\lambda(x,y)$ are well known, see Berezanskij [8].

Now we give an interpretation of Theorem 18.

Theorem 33. Let the operator A be self-adjoint in $L_2(\Omega)$. Let be $0 < \Theta < 1$. Let r be the number with $m_r < 2m\Theta - \frac{1}{2} \leq m_{r+1}$. Then

$$D(A^\Theta) = \Big\{ f \mid f \in W_2^{2m\Theta}(\Omega)\ ,\ B_j f\big|_{\partial\Omega} = 0 \quad \text{for} \quad j = 1, \ldots, r\ ;$$
$$\int_\Omega \frac{1}{d(x)} |(B_{r+1}f)(x)|^2\,dx < \infty \quad \text{if} \quad m_{r+1} = 2m\Theta - \frac{1}{2} \Big\}\ .$$

(The coefficients $b_{r+1,\beta}(x)$ are extended to functions of $C^\infty(\overline{\Omega})$ in an arbitrary way.)

This follows immediately from Theorem 18 and

$$(110)\qquad D(A^\Theta) = (H, D(A))_{\Theta,2}\ ;\quad 0 < \Theta < 1$$

for a self-adjoint operator A in a Hilbert space H.

The theory developed in 2.1.2 enables us to determine the spaces $D(A^\varkappa)$; $\varkappa \geq 0$. The method is clear, hence we do not go into details.

2.2. General differential operators of Legendre type

Let be $-\infty < a < b < \infty$. Let $p(x)$ be the weight function (85) and let $H^{\mu}_{\varkappa_1,\varkappa_2}$ be the spaces (86). We consider the operator A,

$$Af = (-1)^m \, (p^k(x) f^{(m)})^{(m)} ,$$

$D(A) = \{C^\infty([a,b])\,,\ f^{(j)}(a) = f^{(j)}(b) = 0$ for $j = 1, \ldots, m-k-1\}$, in $L_2((a,b))$. Here $k = 0, 1, \ldots, 2m-1$. (For $m-k-1 < 0$ it is $D(A) = C^\infty([a,b])$.) Operators of such type are called general operators of Legendre type. For $q(x) = (x-a)(b-x)$ and $k = m = 1$ we have the usual Legendre operator. The theory of these operators has been developed in [59, 60].

Theorem 34. (a) The closure $\overline{A}$ of A is self-adjoint in $L_2((a,b))$.

(b) For $1 \leqq k \leqq m$ it is $D(\overline{A}^\infty) = \bigcap_{j=1}^{\infty} D(\overline{A}^j) \subset C^\infty([a,b])$. $D(\overline{A}^{j/2})$ is the completion of $D(\overline{A}^\infty)$ in $H^{jm}_{jk,0}$; $j = 0, 1, 2, \ldots$.

A proof of this theorem is given in [59, 60]. In the special case $k = m$ it is $D(\overline{A}^\infty) = C^\infty([a,b])$. Then it follows

$$D(\overline{A}^{j/2}) = H^{jm}_{jk,0} .$$

In [59] it is shown that Theorem 34 (b) is not true for $k > m$. The main purpose of [60] is a discussion of the domains of the fractional powers of $\overline{A}$. The basis of the discussion is the interpolation results of 1.5.2. We succeeded in obtaining a final result only for the interesting case $k = m = 1$.

Theorem 35. Let be $k = m = 1$. Then

(111) $\quad D(\overline{A}^\varkappa) = H^{2\varkappa}_{2\varkappa,0}$ for $\varkappa > \frac{1}{2}$ and $\varkappa \neq \frac{3}{4}, \frac{5}{4}, \frac{7}{4}, \ldots$;

(112) $\quad D(\overline{A}^\varkappa) = \overset{\circ}{H}{}^{2\varkappa}_{2\varkappa,0}$ for $0 \leqq \varkappa \leqq \frac{1}{2}$ and $\varkappa \neq \frac{1}{4}$;

$D(\overline{A}^\varkappa)$ is the completion of $C^\infty([a,b])$ in the norm

(113) $$\left(\|u\|^2_{H^{2\varkappa}_{2\varkappa,0}} + \int_a^b p^{\ell-1}(x)\,|u^{(\ell)}|^2\,dx\right)^{\frac{1}{2}}$$

for $\varkappa = \frac{1}{4} + \frac{\ell}{2}$; $\ell = 1, 2, \ldots$.

$D(\bar{A}^{1/4})$ is the completion of $C_0^\infty((a,b))$ in the norm (113) with $\ell = 0$.

A proof is given in [60]. Nevertheless we have to add a remark. In [60] we formulated the result in the form (111), (112) also for the "singular" values $\varkappa = \frac{1}{4} + \frac{\ell}{2}$; $\ell = 0, 1, 2, \ldots$. We have to correct our result in the given way. The proof is the same.

In the general case we have not been able to obtain a final result. We formulate a partial result.

Theorem 36. Let be $0 < k < m$; $k \equiv 0(2)$; $(k,m) = 1$.

(a) For $\varkappa \geq 0$; $2m\varkappa \neq \frac{1}{2}, \frac{3}{2}, \frac{5}{2}, \ldots$; $(4m - 2k)\varkappa \neq 1, 3, 5, \ldots$, $D(\bar{A}^\varkappa)$ is the completion of $D(\bar{A}^\infty)$ in $H^{2m\varkappa}_{2k\varkappa,0}$.

(b) For $\lambda < 0$ it is $\lambda \notin S_{\bar{A}}$. For $\lambda < 0$ it is further

$$[(\bar{A} - \lambda E)^{-1} f](x) = \int_a^b G_\lambda(x,y) f(y)\, dy$$

with

$$(114) \quad G_\lambda(x,y) \in H^{2m-1/2-\varepsilon}_{2k-k/(2m)-\varepsilon k/m,0} \hat{\otimes} L_2 \cap L_2 \hat{\otimes} H^{2m-1/2-\varepsilon}_{2k-k/(2m)-\varepsilon k/m,0}$$

for each $\varepsilon > 0$. For $\varepsilon < 0$ $G_\lambda(x,y)$ does not belong to the space of the right-hand side of (114).

The part (a) is proved in [60]. We have also to do a slight correction for $2m\varkappa = \frac{1}{2}, \frac{3}{2}, \frac{5}{2}, \ldots$. The part (b) follows immediately from the eigenvalue distribution for the operator $\bar{A}$, see [59], and the method developed in [55].

Finally we formulate an interesting result which follows from a theorem of Baouendi and Goulaouic [6, 7] and the interpolation Theorem 23. We consider the spaces $W^m_{2,\sigma}(\Omega)$ and $B^s_{2,q,\sigma}(\Omega)$ determined in (76) and (82) with $\sigma(x) = p(x)$, $\Omega = (a,b)$; $m = 1, 2, \ldots$; $s > 0$; $1 \leq q \leq \infty$. We set $W^s_{2,p}((a,b)) = B^s_{2,2,p}((a,b))$ for $s > 0$; $s \neq$ integer .

Theorem 37. Let be $k = m = 1$ and $\lambda \notin [0,\infty)$. Then $A - \lambda E$ is an isomorphic map from $W^{s+2}_{2,p}((a,b))$ onto $W^s_2((a,b))$ and from $B^{s+2}_{2,q,p}((a,b))$ onto $B^s_{2,q}((a,b))$. Here $1 \leq q \leq \infty$; $s \geq 0$ for the W-spaces; $s > 0$ for the B-spaces.

For the W-spaces with $s = 0, 1, 2, \ldots$ the theorem is due to Baouendi and Goulaouic [7] (it is not difficult to see that the spaces used by Baouendi and Goulaouic are the spaces $W^j_{2,p}((a,b))$.). The general case follows from Theorem 23 and Theorem 16. It seems to be possible to extend the theorem to the case $k = m$. It would be interesting to develop an L_p-theory for the general Legendre operators. But this seems to be somewhat complicated.

2.3. General differential operators of Tricomi type

The generalization of differential operators of Legendre type to several variables are the so-called operators of Tricomi type. We assume that Ω is a bounded domain with $\partial\Omega \in C^\infty$. There are several possibilities of generalization. For instance for an elliptic operator of the second order

$$Au = \sum_{i,j=1}^{n} a_{ij}(x) \frac{\partial^2 u}{\partial x_i \partial x_j} + \sum_{i=1}^{n} b_i(x) \frac{\partial u}{\partial x_i} + c(x)u ,$$

$a_{ij}(x)$, $b_j(x)$, $c(x) \in C^\infty(\bar{\Omega})$; $a_{ij}(x)$ real functions, we can consider the matrix $(a_{i,j}(x))_{1 \leq i,j \leq n}$, $x \in \Omega$. Let $\lambda_j(x)$ be the positive eigenvalues of this matrix with

$$0 < \lambda_1(x) \leq \lambda_2(x) \leq \ldots \leq \lambda_n(x) .$$

There are two possibilities for the behaviour of $\lambda_j(x)$ as a generalization of Legendre operators.

1. $\exists c_1 > 0$, $c_2 > 0$ with

$$c_1 d(x) \leq \lambda_1(x) \leq c_2 d(x) \tag{115}$$

($d(x)$ is the distance of $x \in \Omega$ from $\partial\Omega$).

2. $\exists c_1 > 0$, $c_2 > 0$ with

$$c_1 d(x) \leq \lambda_1(x) \leq \lambda_n(x) \leq c_2 d(x) . \tag{116}$$

Operators of the second type are considered by Baouendi and Goulaouic [6, 7] and operators of the first type are considered in [56, 58]. When generalizing to operators of even order $2m$ we restrict our attention to characteristic operators:

1. A prototype of an operator of the second type is

$$Af = (-1)^m \sum_{|\alpha|=m} D^\alpha(\varrho^k(x) D^\alpha f) \; ; \quad k = 1, \ldots, 2m-1 , \tag{117}$$

$0 < \varrho(x) \in C^\infty(\Omega)$, $\varrho(x) = d(x)$ near the boundary.

2. A prototype of an operator of the first type may be described in the following way. Near the boundary $\partial\Omega$ we can introduce a local system of coordinates $(y_1,\ldots,y_{n-1},y_n)$, where $(y_1,\ldots,y_{n-1})$ is a general point of $\partial\Omega$ and y_n has the direction of the inner normal ν. In this sense it is

$$S = \partial\Omega \times (0,h) \subset \Omega \; ; \quad h \leq h_0$$

and

$$L_2(S) = L_2(\partial\Omega) \hat{\otimes} L_2((0,h)) \; . \tag{118}$$

Let B be a regular elliptic symmetric positive differential operator of order $2m$ on the C^∞-manifold $\partial\Omega$ (for instance the operator Δ^m where Δ is the second Beltrami operator on $\partial\Omega$). Then it is possible to construct symmetric positive operators A which are regular elliptic in $\Omega - S$ and coincide in S with

$$(-1)^m \frac{\partial^m}{\partial y_n^m}\left(d^k(x) \frac{\partial^m f}{\partial y_n^m}\right) \otimes E + E \otimes B \; ; \quad k = 1, \ldots, 2m-1 \; . \tag{119}$$

Operators of such type are considered in [58]. They are helpful for the construction of nuclear function spaces.

Theorem 38. (a) Let

$$Af = \sum_{|\alpha|\leq 2m} a_\alpha(x) D^\alpha f$$

be a symmetric positive (in $L_2(\Omega)$) operator, regular elliptic in $\Omega - S$, coinciding with (119) in S, with the domain of definition

$$D(A) = \left\{ f \mid f \in C^\infty(\overline{\Omega}) \, , \; \frac{\partial^j f}{\partial \nu^j}\Big|_{\partial\Omega} = 0 \;\text{ for }\; j = 0, \ldots, m-k-1 \right\}$$

(for $m-k-1 < 0$ that means $D(A) = C^\infty(\overline{\Omega})$). Then $\overline{A}$ (as an operator in $L_2(\Omega)$) is self-adjoint.

(b) For $1 \leq k \leq m$ it is $D(\overline{A}^\infty) = \bigcap_{j=1}^{\infty} D(\overline{A}^j) \subset C^\infty(\overline{\Omega})$. For $k = m$ it is $D(\overline{A}^\infty) = C^\infty(\overline{\Omega})$. $D(\overline{A}^{j/2})$ is the completion of $D(\overline{A}^\infty)$ in the norm

$$\left[\|f\|^2_{W_2^{jm}(\Omega - S)} + \int_S \left(\sum_{\ell=1}^{n-1} \left|\frac{\partial^{jm} f}{\partial y_\ell^{jm}}\right|^2 + d^{jk}(x) \left|\frac{\partial^{jm} f}{\partial y_n^{jm}}\right|^2 \right) dx + \|f\|^2_{L_2(\Omega)} \right]^{\frac{1}{2}}$$

($y_1, \ldots, y_n$ are the local coordinates described above, $j = 0, 1, 2, \ldots$.)

This theorem is a generalization of Theorem 34, see [58]. In [58] a more detailed description of $D(\bar{A}^{\infty})$ is also given. By means of Theorem 35 and Theorem 36 it is also possible to give a description of $D(\bar{A}^{\varkappa})$. The method is clear, so we do not go into details.

It would be interesting to give an L_p-theory for operators of such type.

Finally we describe a generalization of Theorem 37. We use the spaces $W^m_{2,\sigma}(\Omega)$ and $B^s_{2,q,\sigma}(\Omega)$ determined in (76) and (82); $m = 1, 2, \ldots$; $s > 0$; $1 \leq q \leq \infty$ with $\sigma(x) = \varrho(x)$. ($\varrho(x)$ is determined in (117).) We set $W^s_{2,\varrho}(\Omega) = B^s_{2,2,\varrho}(\Omega)$ for $s > 0$; $s \neq$ integer.

Theorem 39. Let be

$$Af = -\sum_{j=1}^{n} \frac{\partial}{\partial x_j}\left(\varrho(x)\,\frac{\partial f}{\partial x_j}\right),$$

$0 < \varrho(x) \in C^{\infty}(\Omega)$, $\varrho(x) = d(x)$ near the boundary. For $\lambda \notin [0,\infty)$, $A - \lambda E$ is an isomorphic map from $W^{s+2}_{2,\varrho}(\Omega)$ onto $W^s_2(\Omega)$ and from $B^{s+2}_{2,q,\varrho}(\Omega)$ onto $B^s_{2,q}(\Omega)$. Here $1 \leq q \leq \infty$; $s \geq 0$ for the W-spaces; $s > 0$ for the B-spaces.

For the W-spaces with $s = 0, 1, 2, \ldots$ the theorem is due to Baouendi and Goulaouic [7]. The general case follows from Theorem 23 and Theorem 16.

It would be interesting to develop a systematic theory of the Legendre operators and Tricomi operators of even order of the described types in the L_p-space (or at least in the L_2-space).

2.4. Strongly singular elliptic differential operators

We consider elliptic differential operators which we call "strongly singular". Let Ω be an arbitrary bounded domain in R_n without any smoothness properties. Let $p(x)$ be an infinitely differentiable real function defined in Ω with:

1. $\exists c > 0$ with

(120) $p(x) \geq c d^{-2}(x)$, ($d(x)$ is the distance of $x \in \Omega$ from $\partial\Omega$).

2. $\exists \sigma$; $0 \leqq \sigma < \frac{1}{2}$ with

$$|D^{\gamma}p| \leqq c(|\gamma|)p^{1+\sigma|\gamma|}(x)$$

for all multiindices γ.

It is clear that for such a function $p(x)$ the conditions A of 1.6.1 are fulfilled. It is easy to give examples of functions of such type. Let Ω be a bounded domain with $\partial\Omega \in C^{\infty}$. Each infinitely differentiable positive function $p(x)$ with

$$p(x) = d^{-\alpha}(x) \text{ near the boundary}, \quad \alpha > 2 ,$$

is a function of the described type.

In $L_2(\Omega)$ we consider the operator A,

(121) $$Af = \sum_{|\alpha| \leqq 2m} a_{\alpha}(x)D^{\alpha}f$$

with:

1. $a_{\alpha}(x) \in C^{\infty}(\bar{\Omega})$ for $|\alpha| = 2m$. The ellipticity conditions 1 and 2 of 2.1.1 are fulfilled.

2. $a_0(x)$ ($= a_{\alpha}(x)$ with $\alpha = (0,\ldots,0)$) is a real C^{∞}-function with

$$c_1p^{m}(x) \leqq a_0(x) \leqq c_2p^{m}(x) , \quad c_1 > 0 , \quad c_2 > 0 .$$

3. $a_{\alpha}(x)$ are C^{∞}- functions for $0 < |\alpha| < 2m$. There exist $c(|\gamma|)$ and $\varrho > \frac{1}{2}$ with

(122) $$|a_{\alpha}(x)| \leqq c(0)p^{m-\varrho|\alpha|}(x) ,$$

$$|D^{\gamma}a_{\alpha}(x)| \leqq c_3(|\gamma|)p^{m-|\alpha|/2+\sigma|\gamma|}(x)$$

for $0 < |\alpha| < 2m$.

Obviously the conditions (122) are fulfilled for $a_{\alpha}(x) \in C^{\infty}(\bar{\Omega})$, $0 < |\alpha| < 2m$. The main part of the operator A is

$$\sum_{|\alpha|=2m} a_{\alpha}(x)D^{\alpha}f + a_0(x)f .$$

The conditions (122) are determined in such a way that they do not affect the considerations. The simplest case is the operator A_0,

(123) $$A_0f = -\Delta f + p(x)f .$$

The theory of the operator A_0 is given in [54, 57]. An improvement and an extension to unbounded domains is due to Müller-Pfeiffer [36,

37]. The general operator A is considered by Langemann both in bounded and unbounded domains [28]. The theory of the operator A is based on the interpolation theory developed in 1.6.2, especially Theorem 27. For the sake of brevity we denote by $K^s_\varkappa(\Omega)$; $s \geqq 0$; $\varkappa \geqq 0$ the completion of $C_0^\infty(\Omega)$ in the norm

$$\|f\|_{K^s_\varkappa} = \left(\sum_{|\alpha|=[s]} \|p^{\varkappa/2} D^\alpha f\|^2_{W_2^{\{s\}}(\Omega)} + \|p^{\varkappa/2+s/2} f\|^2_{L_2(\Omega)} \right)^{\frac{1}{2}}, \tag{124}$$

see (92).

Theorem 40. (a) The closure of the operator A determined in (121) with $D(A) = C_0^\infty(\Omega)$, considered as an unbounded operator in $L_2(\Omega)$, has a discrete spectrum.

(b) For $\lambda \notin S_{\overline{A}}$, $A - \lambda E$ is an isomorphic map from $K^{s+2m}_\varkappa(\Omega)$ onto $K^s_\varkappa(\Omega)$ for $s \geqq 0$ and $\varkappa \geqq 0$.

For the operator A_0 , see (123), the theorem is proved in [57]. The general result is due to Langemann [28]. The last theorem gives the possibility to determine $D(A^\theta)$ in the case of a self-adjoint A . Further we can determine the smoothness properties for Green's functions. For this purpose we introduce $K^s_\varkappa(\Omega \times \Omega)$ by (124) after replacing $p(x)$ by $p(x) + p(y)$.

Theorem 41. (a) Let A , determined in (121), be symmetric, $D(A) = C_0^\infty(\Omega)$. The closure $\overline{A}$ of A , considered as an unbounded operator in $L_2(\Omega)$ is self-adjoint. It is

$$D(\overline{A}^\theta) = K_0^{2m\theta}(\Omega), \quad 0 \leqq \theta < \infty .$$

(b) Let A be the operator considered in Theorem 40 (a). Let be $\lambda \notin S_{\overline{A}}$. Then

$$[(A - \lambda E)^{-1} f](x) = \int_\Omega G_\lambda(x,y) f(y)\, dy$$

with

$$G_\lambda(x,y) \in K_0^\varrho(\Omega \times \Omega)$$

iff

$$2m - \frac{n}{2} > \varrho \geqq 0 .$$

A proof is given in [28]. In [28], the theory of the operators A in unbounded domains is also developed. By means of the estimate technique described in [54, 57, 37, 28] it seems to be possible to develop an L_p-theory for these operators. For a further generalization of the results of [54, 57] we refer also to Kniepert [24].

REFERENCES

[1] S. Agmon: On the eigenfunctions and on the eigenvalues of general elliptic boundary value problems. Comm. Pure Appl. Math. 15 (1962), 119 - 147.

[2] S. Agmon: Lectures on elliptic boundary value problems. Van Nostrand Comp., New York 1965.

[3] S. Agmon: Asymptotic formulas with remainder estimates for eigenvalues of elliptic operators. Arch. Rat. Mech. Analysis 28 (1968), 165 - 183.

[4] S. Agmon, A. Douglis, L. Nirenberg: Estimates near the boundary for solutions of elliptic partial differential equations satisfying general boundary conditions I. Comm. Pure Appl. Math. 12 (1959), 623 - 727.

[5] N. Aroszajn, K. T. Smith: Theory of Bessel potentials I. Ann. Inst. Fourier 11 (1961), 385 - 476.

[6] M. S. Baouendi, C. Goulaouic: Étude de la régularité et du spectre d'une classe d'opérateurs elliptiques dégénérés. Compt. Rend. Acad. Sci. Paris Sér. A-B 266 (1968), A336 - A338.

[7] M. S. Baouendi, C. Goulaouic: Régularité et théorie spectrale pour une classe d'opérateurs elliptiques dégénérés. Arch. Rat. Mech. Analysis 34 (1969), 361 - 379.

[8] Ju. M. Berezanskij: Expansions in eigenfunctions of selfadjoint operators (Russian). Naukova Dumka, Kiev 1965.

[9] O. V. Besov: Investigation of a class of function spaces in connection with imbedding and extension theorems (Russian). Trudy Mat. Inst. Steklov. 60 (1961), 42 - 81.

[10] O. V. Besov: Conditions for existence of the classical solution of the wave equation (Russian). Sibirsk. Mat. Ž. 8 (1967), 243 - 256.

[11] O. V. Besov, J. Kadlec, A. Kufner: Certain properties of weight classes (Russian). Dokl. Akad. Nauk SSSR 171 (1966), 514 - 516.

[12] F. E. Browder: On the spectral theory of strongly elliptic differential operators. Proc. Nat. Acad. Sci. U.S.A. 45 (1959), 1423 - 1431.

[13] P. L. Butzer, H. Berens: Semi-groups of operators and approximation. Springer, Berlin 1967.

[14] A. P. Calderón: Intermediate spaces and interpolation. Studia Math., Special Series 1 (1963), 31 - 34.

[15] A. P. Calderón: Intermediate spaces and interpolation, the complex method. Studia Math. 24 (1964), 113 - 190.

[16] S. Fučík, O. John, J. Nečas: On the existence of Schauder bases in Sobolev spaces.

[17] D. Fujiwara: On the asymptotic behaviour of the Green operators for elliptic boundary problems and the pure imaginary powers of some second order operators. Journ. Math. Soc. Japan 21 (1969), 481 - 522.

[18] G. Geymonat, P. Grisvard: Problemi ai limiti lineari ellittici negli spazi di Sobolev con peso. Le Matematiche 22 (1967), 1 - 38.

[19] G. Geymonat, P. Grisvard: Alcuni risultati di teoria spettrale per i problemi ai limiti lineari ellittici. Rendiconti Sem. Matem. Univ. Padova 38 (1967), 121 - 173.

[20] P. Grisvard: Espaces intermediaires entre espaces de Sobolev avec poids. Ann. Scuola Norm. Sup. Pisa 17 (1963), 255 - 296.

[21] P. Grisvard: Commutativité de deux foncteurs d´interpolation et applications. Journ. Math. pures et appl. 45 (1966), 143 - 290.

[22] P. Grisvard: Caractérisation de quelques espaces d´interpolation. Arch. Rat. Mech. Analysis 25 (1967), 40 - 63.

[23] V. P. Il´in: Properties of certain classes of differentiable functions of several variables defined in an n-dimensional domain (Russian). Trudy Mat. Inst. Steklov. 66 (1962), 227 - 363.

[24] D. Kniepert: Generalization of a theorem due to H. Triebel concerning decay properties of eigenfunctions of singular elliptic differential operators. Arch. Rat. Mech. Analysis 37 (1970), 61 - 72.

[25] M. A. Krasnosel´skij, P. P. Zabrejko, E. I. Pustyl´nik, P. E. Sobolevskij: Integral operators in spaces of summable functions (Russian). Nauka, Moskva 1966.

[26] S. G. Krejn, Ju. I. Petunin: Scales of Banach spaces (Russian). Uspehi Mat. Nauk 21 (1966), 89 - 168.

[27] L. D. Kudrjavcev: Imbedding theorems for functions defined on unbounded regions. Dokl. Akad. Nauk SSSR 153 (1963), 530 - 532. (Russian.)

[28] B. Langemann: Über Differenzierbarkeitseigenschaften der Greenschen Funktionen elliptischer Differentialoperatoren und die Existenz von Lösungen quasilinearer elliptischer Differentialgleichungen in Sobolev-Besov-Räumen mit Gewichtsfunktionen. Dissertation, Rostock 1969.

[29] J. L. Lions, E. Magenes: Problemi ai limiti non omogenei III. Ann. Scuola Norm. Sup. Pisa 15 (1961), 41 - 103.

[30] J. L. Lions, E. Magenes: Problèmes aux limites non homogènes IV. Ann. Scuola Norm. Sup. Pisa 15 (1961), 311 - 326.

[31] J. L. Lions, E. Magenes: Problemi ai limiti non omogenei V. Ann. Scuola Norm. Sup. Pisa 16 (1962), 1 - 44.

[32] J. L. Lions, E. Magenes: Problèmes aux limites non homogènes VI. Journ. d´Analyse Math. 11 (1963), 265 - 288.

[33] J. L. Lions, E. Magenes: Problèmes aux limites non homogènes et applications. Dunod, Paris 1968.

[34] J. L. Lions, J. Peetre: Sur une classe d´éspaces d´interpolation. Inst. Hautes Études Sci. Publ. Math. 19 (1964), 5 - 68.

[35] E. Magenes: Interpolation spaces and partial differential equations (Russian). Uspehi Mat. Nauk 21 (1966), 169 - 218.

[36] E. Müller-Pfeiffer: Differenzierbarkeitseigenschaften verallgemeinerter Lösungen der Schrödingergleichung. Wiss. Zeitschr. Päd. Hochschule Erfurt-Mühlhausen 5 (1969), 19 - 20.

[37] E. Müller-Pfeiffer: Zur Theorie elliptischer und hypoelliptischer Differentialoperatoren. Habilitationsschrift, Jena 1967.

[38] T. Muramatu: On Besov spaces of functions defined in general regions. Publ. Res. Inst. Math. Sci. Kyoto Univ. 6 (1970/71), 515 - 543.

[39] J. Nečas: Les méthodes directes en théorie des équations elliptiques. Prague 1967.

[40] Ju. S. Nikol´skij: Some imbedding theorems for weight spaces and their applications (Russian). Imbedding theorems and their applications (Proc. of the Symposium on imbedding theorems, Baku 1966), Moskva 1970, 185 - 192.

[41] S. M. Nikol´skij: Approximation of functions of several variables and imbedding theorems (Russian). Moskva 1969.

[42] J. Peetre: Sur le nombre de paramètres dans la définition de certain espaces d´interpolation. Ricerche Mat. 12 (1963), 248 - 261.

[43] J. Peetre: Espaces d´interpolation et théorème de Soboleff. Ann. Inst. Fourier 16 (1966), 279 - 317.

[44] J. Peetre: Funderingar om Besov-rum. Unpublished lecture note, Lund 1966.

[45] J. Peetre: Sur les espaces de Besov. Compt. Rend. Acad. Sci. Paris Ser. A 264 (1967), 281 - 283.

[46] J. Peetre: A new approach in interpolation spaces. Studia Math. 34 (1970), 23 - 42.

[47] M. Schechter: General boundary value problems for elliptic equations. Comm. Pure Appl. Math. 12 (1959), 457 - 486.

[48] M. Schechter: Complex interpolation. Composito Math. 18 (1967), 117 - 147.

[49] M. I. Slobodeckij: Generalized Sobolev spaces and their applications. Uč. Zap. LGPU 197 (1958), 54 - 112. (Russian.)

[50] S. L. Sobolev: On a theorem in functional analysis. (Russian.) Mat. Sbornik 4(46) (1938), 471 - 497.

[51] S. L. Sobolev: Some applications of functional analysis in mathematical physics (Russian). Izdat. Leningrad. Gos. Univ., Leningrad 1950.

[52] M. H. Taibleson: On the theory of Lipschitz spaces of distributions on Euclidean n-space I. Journ. Math. Mech. 13 (1964), 407 - 479.

[53] M. H. Taibleson: On the theory of Lipschitz spaces of distributions on Euclidean n-space II. Journ. Math. Mech. 14 (1965), 821 - 839.

[54] H. Triebel: Erzeugung nuklearer lokalkonvexer Räume durch singuläre Differentialoperatoren zweiter Ordnung. Math. Annalen 174 (1967), 163 - 176.

[55] H. Triebel: Eigenschaften Greenscher Funktionen nichtselbstadjungierter allgemeiner elliptischer Operatoren. Studia Math. 30 (1968), 339 - 353.

[56] H. Triebel: Erzeugung des nuklearen lokalkonvexen Raumes $C^{\infty}(\bar{\Omega})$ durch einen elliptischen Differentialoperator zweiter Ordnung. Math. Annalen 177 (1968), 247 - 264.

[57] H. Triebel: Singuläre elliptische Differentialgleichungen und Interpolationssätze für Sobolev-Slobodeckij-Räume mit Gewichtsfunktionen. Arch. Rat. Mech. Analysis 32 (1969), 113 - 134.

[58] H. Triebel: Nukleare Funktionenräume und singuläre elliptische Differentialoperatoren. Studia Math. 38 (1970), 283 - 311.

[59] H. Triebel: Allgemeine Legendresche Differentialoperatoren I. Journ. Funct. Analysis 6 (1970), 1- 25.

[60] H. Triebel: Allgemeine Legendresche Differentialoperatoren II. Ann. Scuola Norm. Sup. Pisa 24 (1970), 1 - 35.

[61] H. Triebel: Spaces of distributions of Besov type on Euclidean n-space. Duality, interpolation.

[62] H. Triebel: On the existence of Schauder bases in Sobolev-Besov spaces. Isomorphic properties.

[63] H. Triebel: Boundary values for Sobolev spaces with weights. Density of $D(\Omega)$ in $W^s_{p,\gamma_0,\ldots,\gamma_r}(\Omega)$ and in $H^s_{p,\gamma_0,\ldots,\gamma_r}(\Omega)$ for $s > 0$ and $r = [s - 1/p]^-$.

[64] H. Triebel: Interpolation theory for function spaces of Besov type defined in domains I.

[65] S. V. Uspenskij: On imbedding theorems in smooth domains. Imbedding theorems and their applications (Proc. of the Symposium on imbedding theorems, Baku 1966), Moskva 1970, 219 - 222.

[66] A. Yoshikawa: Remarks on the theory of interpolation spaces. Journ. Fac. Sci. Univ. Tokyo 15 (1968), 209 - 251.

Friedrich-Schiller-Universität, Sektion Mathematik, Helmholtzweg 1, 69 Jena, GDR

QUELQUES QUESTIONS DE L'ANALYSE MATHEMATIQUE DANS LES ESPACES LOCALEMENT LINEAIRES

ELIZBAR S. TSITLANADZE, TBILISI (URSS)

INTRODUCTION

Un grand nombre de problèmes de l´analyse mathématique conduit à l´étude des espaces fonctionnels métriques localement linéaires. Par exemple, dans le calcul variationnel classique tous les espaces fonctionnels considerés des courbes sont localement linéaires. Dans le but d´étudier les propriétés topologiques des espaces fonctionnels localement linéaires M. A. Lavrent´ev et L. A. Ljusternik ont introduit en 1935 [1] la notion d´application localement isométrique. On sait que les applications isométriques des espaces métriques conservent la distance entre les éléments correspondants des espaces. Dans les espaces fonctionnels localement linéaires l´analogie locale de l´application isométrique est l´isométrie dans un voisinage infiniment petit - l´isométrie locale.

Dans ce travail nous introduisons la notion de topologie faible, nous cherchons des conditions nécessaires et suffisantes de la continuité localement forte des fonctionnelles et nous étudions quelques questions concernant les fonctionnelles dérivables et les opérateurs dans les espaces fonctionnels localement linéaires à dimension infinie. Leur rôle est très important pour l´élaboration du calcul différentiel dans les espaces localement linéaires et pour l´étude des problèmes du calcul variationnel en général.

§1. ESPACES LOCALEMENT LINEAIRES

Soit Y un espace fonctionnel métrique d'éléments y. Supposons qu'il existe un espace E linéaire complet réflexif et normé, vérifiant les conditions suivantes:

1. A chaque élément $y_1 \in Y$, appartenant à un ε_1-voisinage $S(y;\varepsilon_1) \subset Y$ de centre y et de rayon ε_1, correspond un élément $h \in E$ - l'image de l'élément y_1 dans E. Réciproquement, à chaque élément $h \in E$, appartenant à un ε_2-voisinage $S(\theta;\varepsilon_2)$ de centre au point neutre θ de E et de rayon ε_2, correspond un élément $y_1 \in Y$ - l'image inverse de l'élément h. Cette correspondance entre les éléments des voisinages $S(y;\varepsilon_1)$ et $S(\theta;\varepsilon_2)$ est topologique, l'élément θ correspondant au point y.

2. Il existe une constante $\mu > 0$ telle que pour tout $y_1 \in S(y;\varepsilon_1)$ on ait l'inégalité

$$\|h\| < \mu\varrho(y,y_1) ,$$

ϱ étant la métrique de l'espace Y.

L'espace E satisfaisant aux conditions 1 et 2 sera appelé espace tangeant à l'espace Y. L'homéomorphisme V, dont l'inverse est l'opérateur V^{-1}, réalisant l'application biunivoque et topologique du voisinage $S(y;\varepsilon_1)$ dans le voisinage $S(\theta;\varepsilon_2)$ et satisfaisant aux conditions 1 et 2, sera appelé application localement isométrique.

On appele espace localement linéaire un espace métrique tel qu'il existe une application topologique d'un voisinage suffisamment petit de chaque point de cet espace sur un voisinage de l'élément neutre de l'espace E, satisfaisant aux conditions 1 et 2.

Théorème 1. Un espace linéaire normé E est un espace localement linéaire dont l'espace tangeant en chacun de ses points est l'espace E lui-même.

Démonstration. Soit $S(y;\varepsilon_1) \subset E$ un ε_1-voisinage d'un point arbitraire $y \in E$, dont les éléments sont $\tilde{y} \in E$, tels que $\varrho(\tilde{y},y) < \varepsilon_1$. Considérons une application univoque et continue par rapport à la norme du voisinage $S(y;\varepsilon_1)$:

$$V\tilde{y} = \tilde{y} - y ; \tag{1}$$

quand $\tilde{y} = y$, on a $Vy = \theta$, où θ est l'élément neutre de l'espace E. Pour $\tilde{y} \neq y$ on a $\|V\tilde{y}\| = \|\tilde{y} - y\| < \varepsilon_1$. Donc, l'application (1) envoie tous les éléments $\tilde{y} \in S(y;\varepsilon_1)$ dans un ε_1-voisinage $S(\theta;\varepsilon_1)$ du point neutre de l'espace E, l'élément $y \in S(y;\varepsilon_1)$ étant envoyé sur θ. Soit h l'image de l'élément $y \in S(y;\varepsilon_1)$ dans $S(\theta;\varepsilon_1)$,

$$h = V\tilde{y} = \tilde{y} - y . \tag{2}$$

On a

$$\|h\| = \|\tilde{y} - y\| < \mu\varrho(\tilde{y},y) ,$$

où $\mu > 1$ est un nombre positif arbitraire et l'inégalité précédente est vraie pour tout $\tilde{y}$. Si l'on définit $\tilde{y}$ de (2), on obtient l'opérateur inverse V^{-1} de V

$$V^{-1}h = \tilde{y} = y + h ,$$

qui est une application univoque et continue de h ; quand $h = \theta$, $V^{-1}\theta = y$, quand $h \neq \theta$, $V^{-1}h = \tilde{y} \in S(y;\varepsilon_1)$. Donc, l'opérateur V^{-1} représente une application du voisinage $S(\theta;\varepsilon_1)$ sur le voisinage $S(y;\varepsilon_1)$ et $V^{-1}\theta = y$. D'où il résulte que les conditions 1 et 2 sont remplies, ce qui achève la démonstration.

§2. LES FONCTIONNELLES LOCALEMENT LINÉAIRES

Soit V une application localement isométrique du voisinage $S(y;\varepsilon_1) \subset Y$ sur le voisinage $S(\theta;\varepsilon_2) \subset E$: $V\tilde{y} = \tilde{x}$, ou $\tilde{x}$ est l'image de $\tilde{y}$ dans $S(\theta;\varepsilon_2)$.

Toute fonctionnelle $f(\tilde{y})$ dans $S(y;\varepsilon_1)$ définit une fonctionnelle $f_1(\tilde{x}) = f(V^{-1}\tilde{x}) = f(\tilde{y})$ dans $S(\theta;\varepsilon_2)$, où V^{-1} est l'inverse de V. Si $f_1(\tilde{x})$ est une fonctionnelle linéaire et continue dans $S(\theta;\varepsilon_2)$, et donc dans E, alors $f(\tilde{y})$ sera appelé fonctionnelle localement linéaire dans $S(y;\varepsilon_1)$ [2]. On obtient des exemples de fonctionnelles localement linéaires si l'on considère les variations dans de différents problèmes classiques des variations.

Théorème 2. Toute fonctionnelle $f(\tilde{y})$ localement linéaire dans le voisinage $S(y;\varepsilon_1) \subset Y$, $\tilde{y} \in S(y;\varepsilon_1)$ est continue dans $S(y;\varepsilon_1)$ par rapport à la métrique de Y.

Démonstration. En effet, soit $y^* \in S(y;\varepsilon_1)$ un point quelconque et soit $\{y_n\} \subset S(y;\varepsilon_1)$ une suite qui converge vers y^* dans la métrique de l'espace Y : $\lim_{n\to\infty} \varrho(y_n, y^*) = 0$. Notons $f_1(\tilde{x})$ la fonctionnelle linéaire définie dans $S(\Theta;\varepsilon_2)$ par le fonctionnelle $f(\tilde{y})$ localement linéaire dans $S(y;\varepsilon_1)$, où $\tilde{x} = V\tilde{y} \in S(\Theta;\varepsilon_2)$.

On a

$$f(y_n) - f(y^*) = f_1(x_n) - f_1(x^*) , \tag{3}$$

où $x^* = Vy^*$, $x_n = Vy_n \in S(\Theta;\varepsilon_2)$. De la continuité de l'homéomorphisme V , on a l'égalité

$$\lim_{n\to\infty} \|Vy_n - Vy^*\| = 0 ,$$

ou bien

$$\lim_{n\to\infty} \|x_n - x^*\| = 0 .$$

La fonctionnelle $f_1(\tilde{x})$ étant continue, on a

$$\lim_{n\to\infty} f_1(x_n) = f_1(x^*)$$

et, en raison de (3), on obtient $\lim_{n\to\infty} f(y_n) = f(y^*)$.

Théorème 3. Si $f(\tilde{y})$ est une fonctionnelle localement linéaire dans $S(y;\varepsilon_1)$ le produit $cf(\tilde{y})$ sera de même localement linéaire dans $S(y;\varepsilon_1)$ pour une constante arbitraire c .

La démonstration est évidante.

Théorème 4. Si $\varphi_i(\tilde{y})$ $(i = 1, 2, \ldots, n)$ sont des fonctionnelles loealement linéaires dans $S(y;\varepsilon_1)$, l'expression linéaire $f(\tilde{y}) = \sum_{i=1}^{n} c_i\varphi_i(\tilde{y})$ sera de même une fonctionnelle localement linéaire dans $S(y;\varepsilon_1)$, où c_i $(i = 1, 2, \ldots, n)$ sont des constantes arbitraires.

Démonstration. En effet, on a

$$f(\tilde{y}) = \sum_{i=1}^{n} c_i\varphi_i(\tilde{y}) = \sum_{i=1}^{n} c_i\varphi_i(V^{-1}\tilde{x}) = \sum_{i=1}^{n} c_i f_i(\tilde{x}) , \tag{4}$$

où $f_i(\tilde{x})$ est une fonctionnelle linéaire dans E , définie par la fonctionnelle $\varphi_i(\tilde{y})$ dans $S(\Theta;\varepsilon_2)$. Le théorème se déduit de l'égalité (4).

§3. LA CONVERGENCE LOCALEMENT FAIBLE

Conservons les notations du paragraphe 2 et introduisons la notion de convergence localement faible dans l´espace Y .

Définition. La suite $\{y_n\} \subset S(y;\varepsilon_1)$ sera appelée localement faiblement convergente vers l´élément $\bar{y} \in S(y;\varepsilon_1)$, que l´on notera $y_n \overset{lf}{\to} \bar{y}$, si pour toute fonctionnelle f localement linéaire dans $S(y;\varepsilon_1)$ on a l´égalité

$$\lim_{n\to\infty} f(y_n) = f(\bar{y}) .$$

L´élément $\bar{y}$ sera appelé la limite localement faible de la suite $\{y_n\}$.

Les propositions suivantes se démontrent aisément:

Théorème 5. Si la suite $\{y_n\} \subset S(y;\varepsilon_1)$ converge localement faiblement vers la limite localement faible $\bar{y} \in S(y;\varepsilon_1)$, la suite des nombres $\{\varrho(y_n,\bar{y})\}$ sera bornée.

Théorème 6. Si la suite $\{y_n\} \subset S(y;\varepsilon_1)$ converge localement faiblement vers la limite localement faible $\bar{y} \in S(y;\varepsilon_1)$, chaque suite $\{y_{n_k}\} \subset \{y_n\}$ converge localement faiblement vers la même limite $\bar{y}$.

Théorème 7. Si la suite $\{y_n\} \subset S(y;\varepsilon_1)$ converge localement faiblement vers la limite localement faible $\bar{y} \in S(y;\varepsilon_1)$ et si la suite $\{y_n'\} \subset S(y;\varepsilon_1)$ vérifie la condition $\lim_{n\to\infty} \varrho(y_n,y_n') = 0$, on aura $y_n' \overset{lf}{\to} \bar{y}$.

Théorème 8. Si la suite $\{y_n\} \subset S(y;\varepsilon_1)$ converge localement faiblement vers la limite localement faible $\bar{y} \in S(y;\varepsilon_1)$ dans l´espace métrique Y localement lineaire, dans l´espace linéaire normé E tangeant à Y , la suite des images $\{x_n\} = \{Vy_n\} \subset S(\theta;\varepsilon_2) \subset E$ convergera faiblement vers la limite faible $\bar{x} = V\bar{y} \in S(\theta;\varepsilon_2)$.

Démonstration. Soit $f_1(\tilde{x})$ une fonctionnelle arbitraire linéaire sur $S(\theta;\varepsilon_2)$, où $\tilde{x}$ est un élément arbitraire de $S(\theta;\varepsilon_2)$. $f_1(\tilde{x})$ est linéaire dans tout l´espace E . $f_1(\tilde{x})$ définit dans $S(y;\varepsilon_1)$ une fonctionnelle $f(\tilde{y})$ localement linéaire: $f_1(\tilde{x}) = f_1(V\tilde{y}) = f_1V(\tilde{y}) = f(\tilde{y})$, où $\tilde{y}$ est l´image inverse dans $S(y;\varepsilon_1)$ de l´élément $\tilde{x}$. On a $\lim_{n\to\infty} |f_1(x_n) - f_1(\bar{x})| = \lim_{n\to\infty} |f(y_n) - f(\bar{y})|$. Or $y_n \overset{lf}{\to} \bar{y}$. Donc, $\lim_{n\to\infty} f(y_n) = f(\bar{y})$. D´où il résulte $x_n \overset{f}{\to} \bar{x}$.

Théorème 9. Si la suite des images $\{x_n\} = \{Vy_n\} \subset S(\theta;\varepsilon_2) \subset E$ converge faiblement vers la limite faible $\bar{x} \in S(\theta;\varepsilon_2)$ dans l'espace linéaire E normé complet et tangeant à Y, la suite des images $\{y_n\} = \{V^{-1}x_n\} \subset S(y;\varepsilon_1) \subset Y$ convergera localement faiblement vers la limite localement faible $\bar{y} = V^{-1}\bar{x} \in S(y;\varepsilon_1)$ dans l'espace métrique Y localement linéaire.

Démonstration. En effet, soit f une fonctionnelle localement linéaire sur $S(y;\varepsilon_1)$. Alors $f_1(x_n) = f(V^{-1}x_n) = f(y_n)$ et $f_1(\bar{x}) = f(V^{-1}\bar{x}) = f(\bar{y})$ sont des fonctionnelles linéaires dans $S(\theta;\varepsilon_2)$. Comme $x_n \overset{f}{\to} \bar{x}$, on a $\lim_{n\to\infty} f_1(x_n) = f_1(\bar{x})$. Donc, $\lim_{n\to\infty} f(y_n) = f(\bar{y})$, ce qui veut dire que $y_n \overset{lf}{\to} \bar{y}$.

Théorème 10. La limite localement faible $\bar{y} \in S(y;\varepsilon_2)$ d'une suite $\{y_n\} \subset S(y;\varepsilon_1)$ localement faiblement convergeante est unique.

Démonstration. Supposons que la suite $\{y_n\} \subset S(y;\varepsilon_1)$ converge localement faiblement vers deux limites localement faibles y_1, $y_2 \in S(y;\varepsilon_1)$. En raison du théorème 8, la suite $\{x_n\} = \{Vy_n\}$ converge faiblement vers les limites faibles $\bar{x}_1 = V\bar{y}_1$, $\bar{x}_2 = V\bar{y}_2 \in S(\theta;\varepsilon_2)$. Or, on sait que la limite faible est unique: $\bar{x}_1 = \bar{x}_2$. Ce qui veut dire $V\bar{y}_1 = V\bar{y}_2$. Mais comme V est un homéomorphisme, il en résulte $\bar{y}_1 = \bar{y}_2$.

Théorème 11. Si la suite $\{y_n\} \subset S(y;\varepsilon_1)$ converge vers $y^* \in S(y;\varepsilon_1)$ par rapport à la métrique de l'espace Y, la suite $\{x_n\} = \{Vy_n\} \subset S(\theta;\varepsilon_2)$ convergera vers $x^* = Vy^* \in S(\theta;\varepsilon_2)$ par rapport à la norme.

La démonstration est évidente.

Théorème 12. Si la suite $\{y_n\} \subset S(y;\varepsilon_1)$ converge vers $y^* \in S(y;\varepsilon_1)$ par rapport à la métrique de l'espace Y alors elle convergera aussi localement faiblement vers y^*.

Démonstration. D'après la condition du théorème, on a $\lim_{n\to\infty} \varrho(y_n,y^*) = 0$. Mais en raison du théorème 11, on a l'égalité: $\lim_{n\to\infty} \|x_n - x^*\| = 0$, où $x^* = Vy^*$, $x_n = Vy_n \subset S(\theta;\varepsilon_2)$. Soit f une fonctionnelle localement linéaire dans $S(y;\varepsilon_1)$ et soit f_1 une fonctionnelle linéaire dans $S(\theta;\varepsilon_2)$ définie par la fonctionnelle f. On a

$$\lim_{n\to\infty} |f(y_n) - f(y^*)| = \lim_{n\to\infty} |f_1(x_n) - f_1(x^*)| .$$

Comme f_1 est une fonctionnelle linéaire et $x_n \Rightarrow x^*$ il en résulte que $\lim_{n\to\infty} f(y_n) = f(y^*)$, où bien $y_n \overset{lf}{\to} y^*$.

Théorème 13. Si l'espace E tangeant à l'espace Y est de dimension finie, la convergence localement faible de la suite $\{y_n\} \subset S(y;\varepsilon_1)$ vers la limite localement faible $\bar{y} \in S(y;\varepsilon_1)$ entraîne la convergence de $\{y_n\}$ vers $\bar{y}$ par rapport à la métrique de l'espace Y .

Démonstration. En raison du théorème 8, la suite des images $\{x_n\} \subset S(\theta;\varepsilon_2)$ converge faiblement vers $\bar{x} = V\bar{y} \in S(\theta;\varepsilon_2)$. Comme E est de dimension finie, $\lim_{n\to\infty} \|x_n - \bar{x}\| = 0$. D'autre part, de la continuité de l'opérateur V^{-1} on a

$$\lim_{n\to\infty} \varrho(V^{-1}x_n, V^{-1}\bar{x}) = 0 ,$$

ou bien

$$\lim_{n\to\infty} \varrho(y_n,\bar{y}) = 0 .$$

Soit Ω l'ensemble de toutes les fonctionnelles localement linéaires sur $S(y;\varepsilon_1)$. L'ensemble $\omega \subset \Omega$ sera appelé partout dense dans Ω , si pour toute fonctionnelle $f \in \Omega$ localement linéaire sur $S(y;\varepsilon_1)$ et pour tout $\varepsilon > 0$ il existe une fonctionnelle $\tilde{f} \in \omega$ localement linéaire tel que pour tout $\tilde{y} \in S(y;\varepsilon_1)$ on ait l'inégalité

$$|f(\tilde{y}) - \tilde{f}(\tilde{y})| < \varepsilon .$$

Nous présentons un critère simple de convergence localement faible d'une suite d'éléments dans Y .

Théorème 14. La condition nécessaire et suffisante pour que la suite $\{y_n\} \subset S(y;\varepsilon_1)$ converge localement faiblement vers la limite localement faible $\bar{y} \in S(y;\varepsilon_1)$ est que pour toute fonctionnelle $\tilde{f} \in \omega$ on ait l'égalité $\lim_{n\to\infty} \tilde{f}(y_n) = \tilde{f}(\bar{y})$.

La condition nécessaire est évidente. Pour démontrer qu'elle est aussi suffisante il faut montrer que pour toute fonctionnelle $f \in \Omega$ localement linéaire dans $S(y;\varepsilon_1)$ on a l'égalité $\lim_{n\to\infty} f(y_n) = f(\bar{y})$. A la fonctionnelle f on peut faire correspondre une fonctionnelle $\tilde{f} \in \omega$ localement linéaire tel que pour $\varepsilon/3 > 0$ et tout $\tilde{y} \in S(y;\varepsilon_1)$ on ait

$$|f(\tilde{y}) - \tilde{f}(\tilde{y})| < \frac{\varepsilon}{3} . \tag{5}$$

D'autre part, on a

$$(6) \qquad |f(y_n) - f(\bar{y})| \leq$$

$$\leq |f(y_n) - \tilde{f}(y_n)| + |\tilde{f}(y_n) - \tilde{f}(\bar{y})| + |\tilde{f}(\bar{y}) - f(\bar{y})| .$$

D'après (5) on peut écrire

$$|f(y_n) - \tilde{f}(y_n)| < \frac{\varepsilon}{3} \quad \text{pour tout} \quad n ,$$

$$|f(\bar{y}) - \tilde{f}(\bar{y})| < \frac{\varepsilon}{3} .$$

Par ailleurs, d'après la condition du théorème, à partir d'un n suffisament grand, on aura $|\tilde{f}(y_n) - \tilde{f}(\bar{y})| < \varepsilon/3$. Donc, à partir d'un n suffisament grand, en vertu de (6), on aura

$$|f(y_n) - f(\bar{y})| < \varepsilon .$$

Soit $\{f^{(n)}(\tilde{y})\}$ une suite de fonctionnelles localement linéaire sur $S(y;\varepsilon_1)$. Nous dirons que $\{f^{(n)}(\tilde{y})\}$ converge dans $S(y;\varepsilon_1)$ s'il existe une fonctionnelle $f(\tilde{y})$, définie sur $S(y;\varepsilon_1)$ telle que $\lim_{n\to\infty} f^{(n)}(\tilde{y}) = f(\tilde{y})$ pour chaque $\tilde{y} \in S(y;\varepsilon_1)$.

<u>Théorème 15</u>. Si la suite $\{f^{(n)}(\tilde{y})\}$ de fonctionnelles localement linéaires converge dans $S(y;\varepsilon_1)$, la fonctionnelle limite $f(\tilde{y})$ est une fonctionnelle localement linéaire sur $S(y;\varepsilon_1)$.

<u>Démonstration</u>. En effet, désignons par $\{f_1^{(n)}(\tilde{x})\}$ la suite de fonctionnelles linéaires dans E , définies respectivement par les fonctionnelles de $\{f^{(n)}(\tilde{y})\}$, où $\tilde{x} = V\tilde{y} \in S(\Theta;\varepsilon_2)$. On a

$$f(\tilde{y}) = \lim_{n\to\infty} f^{(n)}(\tilde{y}) = \lim_{n\to\infty} f_1^{(n)}(\tilde{x}) .$$

Il est évident que $\{f_1^{(n)}(\tilde{x})\}$ est une suite convergente de fonctionnelles linéaire. D'après le théorème connu de Banach, la fonctionnelle limite de cette suite sera une fonctionnelle linéaire dans E . Désignons la par $f_1(\tilde{x})$:

$$\lim_{n\to\infty} f_1^{(n)}(\tilde{x}) = f_1(\tilde{x}) .$$

Comme $f_1(\tilde{x}) = f(\tilde{y})$, $f(\tilde{y})$ est une fonctionnelle localement linéaire sur $S(y;\varepsilon_1)$.

§4. APPROXIMATION DES FONCTIONNELLES LOCALEMENT FORTEMENT CONTINUES

On supposera que l'espace linéaire E tangeant à Y et son espace adjoint E^* possèdent des bases biorthogonales dénombrables $\{e_i\} \subset E$, $\{\ell_i\} \subset E^*$:

$$(7) \qquad \ell_i(e_j) = 1 , \quad i = j , \qquad = 0 , \quad i \neq j \quad (i, j = 1, 2, \ldots) .$$

D'autre part, supposons que E est réflexif. Alors E sera un espace de Banach séparable faiblement complet avec une sphère faiblement compacte. Dans ces conditions tout élément $x \in E$ peut s'exprimer de la manière suivante

$$(8) \qquad x = \sum_{i=1}^{\infty} \ell_i(x)e_i$$

avec

$$(9) \qquad \lim_{n\to\infty} \left\| x - \sum_{i=1}^{n} \ell_i(x)e_i \right\| = 0 .$$

De même, tout élément $\ell \in E^*$ s'exprime au moyen de la série

$$\ell = \sum_{i=1}^{\infty} e_i(\ell)\ell_i$$

avec

$$\lim_{n\to\infty} \left\| \ell - \sum_{i=1}^{n} e_i(\ell)\ell_i \right\| = 0 .$$

Notons

$$(10) \qquad A_n x = \sum_{i=1}^{n} \ell_i(x)e_i , \quad R_n x = \sum_{i=n+1}^{\infty} \ell_i(x)e_i ,$$

$$(11) \qquad A_n \ell = \sum_{i=1}^{n} e_i(\ell)\ell_i , \quad R_n \ell = \sum_{i=n+1}^{\infty} e_i(\ell)\ell_i .$$

On aura

$$x = A_n x + R_n x , \quad \ell = A_n \ell + R_n \ell .$$

Donc, les espaces E et E^* se décomposent en sommes directes

$$(12) \qquad E = E_{A_n} + E_{R_n} , \quad E^* = E^*_{A_n} + E^*_{R_n}$$

de sous-espaces linéaires à n dimensions E_{A_n} , $E^*_{A_n}$, respectivement avec bases $\{e_i\}_{i=1}^n$, $\{\ell_i\}_{i=1}^n$ et éléments $A_n x \in E_{A_n}$, $A_n \ell \in E^*_{A_n}$, et de sous-espaces linéaires de dimension infinie E_{R_n} et $E^*_{R_n}$ avec bases $\{e_i\}_{i=n+1}^\infty$, $\{\ell_i\}_{i=n+1}^\infty$ et éléments $R_n x \in E_{R_n}$, $R_n \ell \in E^*_{R_n}$, où $E^*_{A_n}$, $E^*_{R_n}$ sont les espaces adjoints de E_{A_n} et E_{R_n} , respectivement.

Il est aisé de montrer que $A_n x$ est orthogonal à $R_n \ell$ et que $R_n x$ est orthogonal à $A_n \ell$:

$$(13) \qquad (A_n x, R_n \ell) = (R_n \ell, A_n x) = 0 , \quad (R_n x, A_n \ell) = (A_n \ell, R_n x) = 0$$

où, par exemple, $(A_n x, R_n \ell)$ désigne le produit scalaire des éléments $A_n x$, $R_n \ell$ et est égal à la valuer de la fonctionnelle linéaire $R_n \ell$ pour l´élément $A_n x$.

Donc, E_{A_n} est constitué par les éléments orthogonaux aux éléments de $E^*_{R_n}$ et réciproquement. E_{R_n} est constitué par les éléments orthogonaux aux éléments de $E^*_{A_n}$ et réciproquement. Ce qui peut s´écrire de la manière suivante:

$$(14) \qquad (E_{A_n}, E^*_{R_n}) = (E^*_{R_n}, E_{A_n}) = 0 ,$$

$$(E_{R_n}, E^*_{A_n}) = (E^*_{A_n}, E_{R_n}) = 0 .$$

Soit $\tilde{y}$ un élément arbitraire de $S(y;\varepsilon_1)$, dont l´image dans $S(\theta;\varepsilon_2)$ sera désignée par $\tilde{x}$. L´élément $\tilde{x}$ se décompose de la manière suivante: $\tilde{x} = A_n \tilde{x} + R_n \tilde{x}$. Notons $A_n \tilde{y} = V^{-1} A_n \tilde{x}$, $R_n \tilde{y} = V^{-1} R_n \tilde{x}$.

<u>Théorème 16</u>. On a $A_n \tilde{y} = V^{-1} A_n \tilde{x}$, $R_n \tilde{y} = V^{-1} R_n \tilde{x} \in S(y;\varepsilon_1)$ pour n suffisament grand.

<u>Démonstration</u>. Evidement, il suffit de démontrer que pour n suffisament grand on a $A_n \tilde{x}$, $R_n \tilde{x} \in S(\theta;\varepsilon_2)$. On peut faire de telle sorte que, choisissant n suffisament grand, en vertu de (9), la norme de l´élément $R_n \tilde{x}$ devienne arbitrairement petite

$$\|R_n \tilde{x}\| \leq \frac{d}{2} < \varepsilon_2 ,$$

où $d > 0$ et donc, $R_n\tilde{x} \in S(\theta;\varepsilon_2)$. Mais alors, de même $A_n\tilde{x} \in S(\theta;\varepsilon_2)$. En effet, supposons $A_n\tilde{x} \bar{\in} S(\theta;\varepsilon_2)$ et $\|A_n\tilde{x}\| = \varepsilon_2 + d$. Les inégalités

$$\|\tilde{x}\| \geq \left| \|A_n\tilde{x}\| - \|R_n\tilde{x}\| \right| \geq \varepsilon_2 + \frac{d}{2}$$

entraînent une contradiction, car $\tilde{x} \in S(\theta;\varepsilon_2)$.

Théorème 17. Si la suite $\{y_k\} \subset S(y;\varepsilon_1)$ converge localement faiblement vers la limite localement faible $\bar{y} \in S(y;\varepsilon_1)$, la suite $\{A_k y_k\} \subset S(y;\varepsilon_1)$ converge localement faiblement vers la limite localement faible $\bar{y}$.

En effet, soit φ une fonctionnelle localement linéaire sur $S(y;\varepsilon_1)$. On a

$$\varphi(A_k y_k) = \varphi(V^{-1}A_k x_k) = \ell(A_k x_k) \tag{15}$$

où ℓ est une fonctionnelle linéaire définie par la fonctionnelle φ dans $S(\theta;\varepsilon_2)$. En vertu du théorème 8, la suite des images $\{x_k\} \subset S(\theta;\varepsilon_2)$ converge vers la limite faible $\bar{x} = V\bar{y} \in S(\theta;\varepsilon_2)$. Mais alors, d'après un théorème connu (6), la suite $\{A_k x_k\} \xrightarrow{f} \bar{x}$. Donc, de (15) on obtient

$$\lim_{k\to\infty} \varphi(A_k y_k) = \lim_{k\to\infty} \ell(A_k x_k) = \varphi(\bar{y}) .$$

Donc, la suite

$$A_k y_k \xrightarrow{lf} \bar{y} .$$

Théorème 18. Si la suite $\{y_k\} \subset S(y;\varepsilon_1)$ est telle que

$$y_k \xrightarrow{lf} \bar{y} \in S(y;\varepsilon_1) ,$$

la suite $\{A_n y_k\}$ converge localement faiblement vers la limite localement faible $A_n\bar{y}$ quand $k \to \infty$.

En effet, si l'on conserve les notations du théorème 17, on aura

$$\lim_{k\to\infty} \varphi(A_n y_k) = \lim_{k\to\infty} \ell(A_n x_k) .$$

Mais il est aisé de voir que

$$\lim_{k\to\infty} \ell(A_n x_k) = \ell(A_n\bar{x}) .$$

Donc, on a

$$\lim_{k\to\infty} \varphi(A_n y_k) = \ell(A_n\bar{x}) = \varphi(A_n\bar{y})$$

ce qui veut dire

$$A_n y_k \xrightarrow{lf} A_n\bar{y} \text{ , quand } k \to \infty .$$

Considérons maintenant une fonctionnelle arbitraire $f(\tilde{y})$, définie sur $S(y;\varepsilon_1)$ et introduisons la définition suivante:

<u>Définition</u>. $f(\tilde{y})$ sera appelé fonctionnelle localement fortement continue au point $\bar{y} \in S(y;\varepsilon_1)$, si quelle pour n'importe suite $\{y_n\}$ localement faiblement convergente vers la limite faible $\bar{y}$ on a l'égalité

$$\lim_{n\to\infty} f(y_n) = f(\bar{y}) .$$

Si $f(\tilde{y})$ possède cette propriété en chaque point du voisinage $S(y;\varepsilon_1)$, $f(\tilde{y})$ est localement fortement continue sur $S(y;\varepsilon_1)$.

Il est évident que toute fonctionnelle $f(\tilde{y})$ localement fortement continue sur $S(y;\varepsilon_1)$ est continue sur $S(y;\varepsilon_1)$ par rapport à la métrique de l'espace Y . En effet, supposons que la suite $\{y_n\} \subset S(y;\varepsilon_1)$ converge par rapport à la métrique vers l'élément $y^* \in S(y;\varepsilon_1)$. Alors, elle converge localement faiblement vers y^* , car on a $\lim_{n\to\infty} \varrho(y_n;y^*) = 0$, d'où $\lim_{n\to\infty} \|x_n - x^*\| = 0$ pour $x_n = Vy_n$, $x^* = Vy^*$; x_n , $x^* \in S(\theta;\varepsilon_2)$, $n = 1, 2, \ldots$. Si maintenant φ est une fonctionnelle arbitraire localement linéaire sur $S(y;\varepsilon_1)$ et si ℓ est la fonctionnelle linéaire sur $S(\theta;\varepsilon_2)$ définie par φ , on aura

$$\lim_{n\to\infty} |\varphi(y_n) - \varphi(y^*)| = \lim_{n\to\infty} |\ell(x_n) - \ell(x^*)| = 0 .$$

Donc, $y_n \xrightarrow{lf} y^*$.

Comme $f(\tilde{y})$ est localement fortement continue sur $S(y;\varepsilon_1)$, on a $\lim_{n\to\infty} f(y_n) = f(y^*)$ et $f(\tilde{y})$ est continue sur $S(y;\varepsilon_1)$ par rapport à la métrique de l'espace Y .

On obtient un exemple de fonctionnelle sur $S(y;\varepsilon_1)$ continue par rapport à la métrique mais qui n'est pas localement fortement continue, si l'on considère la distance $\varrho(\tilde{y},y)$, où $\tilde{y}$ est un point arbitraire de $S(y;\varepsilon_1)$.

Maintenant nous formulons et démontrons un des théorèmes principaux.

Théorème 19. La condition nécessaire et suffisante pour qu une fonctionnelle continue $f(\tilde{y})$ soit localement fortement continue sur $S(y;\varepsilon_1)$ est que pour chaque $\varepsilon > 0$ il existe un nombre $N = N(\varepsilon)$ tel que pour $n > N$ et $\tilde{y} \in S(y;\varepsilon_1)$ on ait l'inégalité

$$|f(\tilde{y}) - f(A_n\tilde{y})| < \varepsilon . \tag{16}$$

Démonstration. Pour montrer que la condition est nécessaire, supposons que $f(\tilde{y})$ est localement fortement continue sur $S(y;\varepsilon_1)$, mais ne vérifie pas l'inégalité (16). Alors il existe une suite croissante non bornée $\{n_k\}$ d'entiers naturels, telle que pour chaque n_k il existe au moins un élément $y_{n_k} \in S(y;\varepsilon_1)$ qui vérifie l'inégalité

$$|f(y_{n_k}) - f(A_{n_k}y_{n_k})| \geqq \varepsilon . \tag{17}$$

La suite $\{y_{n_k}\}$ définit dans $S(\theta;\varepsilon_2)$ une suite $\{x_{n_k}\} = \{Vy_{n_k}\} \subset S(\theta;\varepsilon_2)$. Comme, dans E, tout ensemble borné par rapport à la norme est faiblement compact, on peut extraire de $\{x_{n_k}\}$ une suite faiblement convergente $\{x_{n_{k_s}}\}$, dont la limite faible sera notée $\bar{x}$. A la suite $\{x_{n_{k_s}}\}$ il correspond dans $S(y;\varepsilon_1)$ une suite $\{y_{n_{k_s}}\} \subset \{y_{n_k}\}$, qui converge localement faiblement vers l'élément $\bar{y} \in S(y;\varepsilon_1)$. En raison de (17), on a l'inégalité

$$|f(y_{n_{k_s}}) - f(A_{n_{k_s}}y_{n_{k_s}})| \geqq \varepsilon . \tag{18}$$

Comme $y_{n_{k_s}} \overset{lf}{\to} \bar{y}$, en vertu du théorème 17, la suite $\{A_{n_{k_s}}y_{n_{k_s}}\}$ converge aussi localement faiblement vers la même limite localement faible: $A_{n_{k_s}}y_{n_{k_s}} \overset{lf}{\to} \bar{y}$.

D'après l'hypothèse du théorème, la fonctionnelle f est localement fortement continue dans $S(y;\varepsilon_1)$. Donc, pour chaque $\varepsilon/2 > 0$ il existe $N = N(\varepsilon/2)$ tel que, quand $n_{k_s} > N$, on ait

$$|f(y_{n_{k_s}}) - f(\bar{y})| < \frac{\varepsilon}{2} , \tag{19}$$

$$|f(\bar{y}) - f(A_{n_{k_s}}y_{n_{k_s}})| < \frac{\varepsilon}{2} .$$

De (19) on obtient aisément

$$|f(y_{n_{k_s}}) - f(A_{n_{k_s}} y_{n_{k_s}})| \leq$$

$$\leq |f(y_{n_{k_s}}) - f(\bar{y})| + |f(\bar{y}) - f(A_{n_{k_s}} y_{n_{k_s}})| < \varepsilon ,$$

ce qui est en contradiction avec l'inégalité (18).

Démontrons maintenant que la condition du théorème est suffisante. Soit $\{y_k\} \subset S(y;\varepsilon_1)$ une suite localement faiblement convergente, dont la limite localement faible est $\bar{y} \in S(y;\varepsilon_1)$.

En vertu de (9), pour la suite des images dans $S(\theta;\varepsilon_2)$ on a

$$\lim_{n\to\infty} \|x_k - A_n x_k\| = 0 , \lim_{n\to\infty} \|\bar{x} - A_n\bar{x}\| = 0 .$$

Mais alors, de la continuité de l'opérateur V^{-1} et de la distance il résulte

$$(20) \qquad \lim_{n\to\infty} \varrho(V^{-1}x_k, V^{-1}A_n x_k) = \lim_{n\to\infty} \varrho(y_k, A_n y_k) = 0 ,$$

$$\lim_{n\to\infty} \varrho(V^{-1}\bar{x}, V^{-1}A_n\bar{x}) = \lim_{n\to\infty} \varrho(\bar{y}, A_n\bar{y}) = 0 .$$

D'autre part, il est évident qu'on peut écrire

$$(21) \qquad |f(y_k) - f(\bar{y})| \leq |f(y_k) - f(A_n y_k)| +$$

$$+ |f(A_n y_k) - f(A_n\bar{y})| + |f(A_n\bar{y}) - f(\bar{y})| .$$

Comme $f(\tilde{y})$ est une fonctionnelle continue sur $S(y;\varepsilon_1)$, en vertu de (20), pour tout $\varepsilon > 0$ il existe $N = N(\varepsilon)$ tel que pour $n > N(\varepsilon)$

$$(22) \qquad |f(y_k) - f(A_n y_k)| < \frac{\varepsilon}{3} ,$$

$$|f(A_n\bar{y}) - f(\bar{y})| < \frac{\varepsilon}{3} .$$

En outre, d'une part la suite $\{A_n x_k\}$ converge faiblement vers la limite faible $A_n\bar{x}$ quand $k \to \infty$ et, d'autre part, les éléments de la suite $\{A_n x_k\}$ et $A_n x$ appartiennent à l'espace E_{A_n} de dimension finie (de n dimensions), dans lequel la convergence faible coincide avec la convergence par rapport à la norme, c'est-à-dire

$$\lim_{k\to\infty} \|A_n x_k - A_n\bar{x}\| = 0 .$$

Il en résulte, quand $k \to \infty$, que la suite $\{A_n y_k\}$ converge vers $A_n \bar{y}$ par rapport à la métrique de l'espace Y :

$$\lim_{k\to\infty} \varrho(A_n y_k, A_n \bar{y}) = 0 .$$

Donc, pour tout $\varepsilon > 0$ il existe un N tel que

$$|f(A_n y_k) - f(A_n \bar{y})| < \frac{\varepsilon}{3}, \quad n > N . \tag{23}$$

En vertu de (22) et (23), on déduit de (21)

$$|f(y_k) - f(\bar{y})| < \varepsilon ,$$

ce qui achève la démonstration du théorème.

Remarque. L'inégalité (16) montre que la valeur de la fonctionnelle $f(\tilde{y})$ localement fortement continue sur $S(y;\varepsilon_1)$ en un point quelconque $\tilde{y} \in S(y;\varepsilon_1)$ peut être approchée pour n suffisamment grand avec une précision donnée par les valeurs de cette fonctionnelle aux éléments de la suite $\{A_n \tilde{y}\}$, le nombre $\varepsilon > 0$ caractérisant le degré de l'approximation obtenu.

BIBLIOGRAPHIE

[1] M. A. Lavrent'ev, L. A. Ljusternik: Les fondements du calcul variationnel (en russe), t. I, par. 2. Moscou 1935, 127 - 129.

[2] L. A. Ljusternik: Intersections dans les espaces linéaires (en russe). DAN URSS 27 (1940), N 8, 771 - 774.

[3] A. G. Sigalov: Les applications presque isométriques et pseudo-dérivation (en russe). DAN URSS 56 (1946), 77 - 78.

[4] L. A. Ljusternik: Les applications presque isométriques (en russe). Uspechi Mat. Nauk 2 (1947), fasc. 1 (17), 218 - 219.

[5] L. A. Ljusternik, V. I. Sobolev: Les éléments de l'analyse fonctionnelle (en russe). Moscou, 1951, 337 - 339.

[6] E. S. Tsitlanadze: Sur la dérivation des fonctionnelles (en russe). Mat. Sb. 29 (71) (1951), 3 - 12.